Springer Theses

Recognizing Outstanding Ph.D. Research

For further volumes:
http://www.springer.com/series/8790

Aims and Scope

The series "Springer Theses" brings together a selection of the very best Ph.D. theses from around the world and across the physical sciences. Nominated and endorsed by two recognized specialists, each published volume has been selected for its scientific excellence and the high impact of its contents for the pertinent field of research. For greater accessibility to non-specialists, the published versions include an extended introduction, as well as a foreword by the student's supervisor explaining the special relevance of the work for the field. As a whole, the series will provide a valuable resource both for newcomers to the research fields described, and for other scientists seeking detailed background information on special questions. Finally, it provides an accredited documentation of the valuable contributions made by today's younger generation of scientists.

Theses are accepted into the series by invited nomination only and must fulfill all of the following criteria

- They must be written in good English.
- The topic should fall within the confines of Chemistry, Physics, Earth Sciences, Engineering and related interdisciplinary fields such as Materials, Nanoscience, Chemical Engineering, Complex Systems and Biophysics.
- The work reported in the thesis must represent a significant scientific advance.
- If the thesis includes previously published material, permission to reproduce this must be gained from the respective copyright holder.
- They must have been examined and passed during the 12 months prior to nomination.
- Each thesis should include a foreword by the supervisor outlining the significance of its content.
- The theses should have a clearly defined structure including an introduction accessible to scientists not expert in that particular field.

Janusz Bogdanowicz

Photomodulated Optical Reflectance

A Fundamental Study Aimed at Non-Destructive Carrier Profiling in Silicon

Doctoral Thesis accepted by
the University of Leuven, Belgium

Springer

Author
Dr. Janusz Bogdanowicz
IMEC
Leuven
Belgium

Supervisor
Prof. Dr. Ir. Wilfried Vandervorst
IMEC
Leuven
Belgium

ISSN 2190-5053 ISSN 2190-5061 (electronic)
ISBN 978-3-642-30107-0 ISBN 978-3-642-30108-7 (eBook)
DOI 10.1007/978-3-642-30108-7
Springer Heidelberg New York Dordrecht London

Library of Congress Control Number: 2012940720

Printed on acid-free paper

Springer is part of Springer Science+Business Media (www.springer.com)

Supervisor's Foreword

The continuous downscaling of nanoelectronic components and transistors following Moore's law leads to ever cheaper, more powerful computers, and more multifunctional devices. To maintain this progress in scaling, material scientists, and process engineers explore new materials, novel processing steps and more complex device structures. These technological developments are extremely challenging and can only be achieved if based on profound physical insight and understanding. As such metrology has become a key enabler toward successful product development and inevitably subject to similar scaling laws in order to remain relevant. Hence research in metrology targeting the extension of measurements toward smaller dimensions, higher resolution, better sensitivity and specificity, and quantification accuracy is an important cornerstone of technological progress in semiconductor technology.

As dopants and dopant distributions constitute the active heart of the transistor operation, engineering of the source and drain junction regions of MOS transistors has become an important research item listed in the International Technology Roadmap for Semiconductors and the subject of a lot of advanced research activities. Their strong impact on the device performance led to a continuous scaling toward very shallow depths and an engineering of very detailed dopant distributions (doping concentration, depth, abruptness, halo,..). Concurrently the development of new doping and annealing processes as well as their production control has become extremely dependent on the availability of precise (electrical) characterization with adequate specifications. In particular the interest in in-line monitoring and non-destructive characterization has spurred the search for junction metrology based on optical probes; however, with the important constraints to provide also sub-nanometer accuracy, precision, and quantification.

This work contributes to this evolution and demonstrates the capabilities of the Photomodulated Optical Reflectance (PMOR) which measures the variations in reflectance induced by a modulated pump laser. Based on a fast laser pump-probe technique, PMOR was originally designed as a non-contact concept for implant dose metrology. However, a breakthrough has been realized in this work by extending the concept from a simple process control tool for as-implant profiles toward the quantitative analysis of a complete active carrier profile through a very detailed modeling of the fundamental physics of the PMOR technique. This work provides a complete description of the carrier and heat generation/transport/recombination phenomena as well as of their translation into reflectance signals with an excellent agreement to experimental data. Novel experimental approaches are introduced to extend PMOR from qualitative dose monitoring toward quantitative junction depth measurement and detailed carrier profile reconstruction. The latter development represents a significant breakthrough in USJ metrology and generated substantial industrial interest, up to the development of a new product implementing the final recommendations of this book.

Leuven, January 2012 Prof. Dr. Ir. Wilfried Vandervorst

Acknowledgments

First and foremost, I would like to thank Prof. Wilfried Vandervorst for giving me the opportunity to carry out my Ph.D. studies under his supervision. I really appreciated his guidance and support throughout these years. Despite his very busy agenda, he always took the time needed for short explanations or long discussions.

I also wish to sincerely acknowledge Trudo Clarysse, who really made the difference all along these years. He was an excellent supervisor who always pushed me to the best of myself. In so many instances, his invaluable assistance and involvement were the key to the results shown in this thesis. I am today fully aware that I owe him much more than I could ever write in these acknowledgments.

The following thanks go to Fabian Dortu, whose help and will to share his deep fundamental and technical understanding gave me a real boost in the early phase of this work. Even 3 years after he left imec, I still have the impression that the numerous technical discussions we used to have paved the way to the final conclusions of this work. I should obviously not forget to mention that he is the original developer of FSEM. Now that this work is reaching an end, I realize that this tool is a real goldmine and I deeply acknowledge him for the energy he put into its development.

I am also indebted to Erik Rosseel, who I very much enjoyed working with. The numerous hours spent discussing technical issues next to the TP tool and all the experimental results collected on his immense experimental database were clear determining factors of this work.

I am very grateful to all the KLA-Thermawave team and more particularly to Derrick Shaughnessy and Alex Salnik. My special thanks to Derrick for the very nice company during my two stays in California.

This work would not have been possible without the help of Duy Nguyen, Laurent Souriau, and all the epi group of imec, who provided me with very nice layers. Thanks also to Alain Moussa, Danielle Vanhaeren, Joris Delmotte, Bastien Douhard, Luc Geenen, Pierre Eyben, Jozefien Goossens, and Benny Van Daele for kindly taking care of the characterization of these layers and to the FPS unit for the measurement time on the ThermaProbe tool.

I certainly will not forget to thank all my (ex-)mca colleagues and friends and most particularly Francesca, Simone, Wouter, Jay, Alain, Fabian, Cindy, Bart, Bastien, Joris, Alexis, and Danielle for the enjoyable time spent together. My imec experience would not have been the same without you.

I would also like to acknowledge the Institute for the Promotion of Innovation through Science and Technology in Flanders (IWT-Vlaanderen) for their financial support.

Last but not least, I would like to thank all the people, family and friends, who I love and who supported me along these years.

Contents

Publications and Conference Contributions

Peer-Reviewed Publications: First Author

Bogdanowicz Janusz, Clarysse Trudo, Gerrit Smets, Rosseel Erik, and Vandervorst Wilfried

Advanced Use of Therma-Probe for Ultra-Shallow Junction Monitoring. Proceedings of the 2011 International Conference on Frontiers of Characterization and Metrology for Nanoelectronics, AIP Conference Proceedings, **1395**: 208, 2011.

Bogdanowicz Janusz, Clarysse Trudo, Moussa Alain, Mody Jay, Eyben Pierre, Vandervorst Wilfried and Rosseel Erik

Non-Destructive Characterization of Activated Ion-Implanted Doping Profiles Based on Photomodulated Optical Reflectance. Proceedings of the 18th International Conference on Ion Implantation Technology, AIP Conference Proceedings, **1321**: 220, 2011.

Bogdanowicz Janusz, Dortu Fabian, Clarysse Trudo, Vandervorst Wilfried and Salnik Alex

Electrothermal Theory Of Photomodulated Optical Reflectance on Active Doping Profiles in Silicon. Journal of Applied Physics, **108**: 104908, 2010.

Bogdanowicz Janusz, Dortu Fabian, Clarysse Trudo, Vandervorst Wilfried, Rosseel Erik, Nguyen Ngoc Duy, Shaughnessy Derrick, Salnik Alex and Nicolaides Lena

Nondestructive Extraction of Junction Depths of Active Doping Profiles from Photomodulated Optical Reflectance Offset Curves. Journal of Vacuum Science & Technology B: Microelectronics and Nanometer Structures, **28(1)**: C1C1, 2010.

Bogdanowicz Janusz, Dortu Fabian, Clarysse Trudo, Vandervorst Wilfried, Shaughnessy Derrick, Salnik Alex, Nicolaides Lena and Opsal Jon

Impact of Inactive Dopants in Chemical Vapor Deposition Layers on Photo-modulated Optical Reflectance. Materials Science & Engineering B, **154–155**: 234–239, 2008.

Bogdanowicz Janusz, Dortu Fabian, Clarysse Trudo, Vandervorst Wilfried, Shaughnessy Derrick, Salnik Alex, Nicolaides Lena and Opsal Jon
Advances in Optical Carrier Profiling through High-Frequency Modulated Optical Reflectance. Journal of Vacuum Science & Technology B: Microelectronics and Nanometer Structures, **26(1)**: 310, 2008.

Peer-Reviewed Publications: Contributions

Smets Gerrit, Rosseel Erik, Sterckx Gunther, Bogdanowicz Janusz, Vandervorst Wilfried and Shaughnessy Derrick
Transfer from Rs-based to PMOR-based Ion Implantation Process Monitoring. Proceedings of the 18th International Conference on Ion Implantation Technology, AIP Conference Proceedings, **1321**: 426, 2011.

Radisic Dunja, Shamiryan Demis, Mannaert Geert, Boullart Werner, Rosseel Erik and Bogdanowicz Janusz
Metrology for Implanted Si Substrate and Dopant Loss Studies. Journal of the Electrochemical Society, **157**: H580, 2010.

Clarysse Trudo, Moussa Alain, Parmentier Brigitte, Bogdanowicz Janusz, Vandervorst Wilfried, Bender Hugo, Pfeffer Markus, Schellenberger Martin, Nielsen Peter F., Thorsteinsson Sune, Lin Rong and Petersen Dirch
Photovoltage versus microprobe sheet resistance measurements on ultrashallow structures. Journal of Vacuum Science & Technology B: Microelectronics and Nanometer Structures, **28(1)**: C1C8, 2010.

Rosseel Erik, Bogdanowicz Janusz, Clarysse Trudo, Vandervorst Wilfried, Ortolland Claude, Hoffmann Thomas, Salnik Alex, Nicolaides Lena and Han Sang-Hyun.
Study of sub-melt laser induced junction non-uniformities using Therma-Probe. Journal of Vacuum Science & Technology B: Microelectronics and Nanometer Structures, **28(1)**: C1C21, 2010.

Clarysse Trudo, Bogdanowicz Janusz, Goossens Jozefien, Moussa Alain, Rosseel Erik, Vandervorst Wilfried, Petersen Dirch, Lind Rong, Nielsen Peter Folmer, Hansen Ole, Merklin Greg, Bennett Nick and Cowern Nick
On the analysis of the activation mechanisms of sub-melt laser anneals. Materials Science & Engineering B, **154–155**: 24–30, 2008.

Fabian Dortu, Janusz Bogdanowicz, Trudo Clarysse, and Wilfried Vandervorst.
Impact of bandgap narrowing and surface recombination on photoelectrothermal modulated optical reflectance (PMOR) power curves. Journal of

Vacuum Science & Technology B: Microelectronics and Nanometer Structures, **26(1)**: 322, 2008.

Fabian Dortu, Janusz Bogdanowicz, Trudo Clarysse, and Wilfried Vandervorst. *Nonlinear study of photoelectrothermal modulated optical reflectance for active dopant profile extraction*. Journal of Applied Physics, **101**: 053107, 2007.

Clarysse Trudo, Eyben Pierre, Janssens Tom, Hoflijk Ilse, Vanhaeren Danielle, Satta Alessandra, Meuris Marc, Vandervorst Wilfried and Bogdanowicz Janusz *Active dopant characterization methodology for germanium*. Journal of Vacuum Science & Technology B: Microelectronics and Nanometer Structures, **24**: 381, 2006.

Non-Peer-Reviewed Publications: Contributions

Rosseel Erik, Petersen Dirch, Osterberg Frederik W., Hansen Ole, Bogdanowicz Janusz, Clarysse Trudo, Vandervorst Wilfried , Ortolland Claude, Hoffmann Thomas, Chan Philip, Salnik Alex and Nicolaides Lena *Monitoring of local and global temperature non-uniformities by means of Therma-Probe and Micro Four-Point Probe metrology*. Proceedings of the 17th IEEE International Conference on Advanced Thermal Processing of Semiconductors, 2009

Rosseel Erik, Bogdanowicz Janusz, Clarysse Trudo, Vandervorst Wilfried, Salnik Alex, Han Sang-Hyun and Nicolaides Lena *Study of sub-melt laser damage annealing using Therma-Probe*. Proceedings of the 17th IEEE International Conference on Advanced Thermal Processing of Semiconductors, 2009

Conference Contributions

Bogdanowicz Janusz, Clarysse Trudo, Gerrit Smets, Rosseel Erik, and Vandervorst Wilfried *Advanced Use of Therma-Probe for Ultra-Shallow Junction Monitoring*. Presented at the 2011 International Conference on Frontiers of Characterization and Metrology for Nanoelectronics (Grenoble, France).

Bogdanowicz Janusz, Clarysse Trudo, Vandervorst Wilfried, Rosseel Erik and Shaughnessy Derrick *Non-Destructive Characterization of Activated Ion-Implanted Doping Profiles Based on Photomodulated Optical Reflectance*. Presented at the 18th International Conference on Ion Implantation Technology (Kyoto, Japan).

Bogdanowicz Janusz, Dortu Fabian, Clarysse Trudo, Vandervorst Wilfried,Shaughnessy Derrick and Salnik Alex

Theory of Photomodulated Optical Reflectance on Active Doping Profiles in Silicon. Presented at the International Conference on Photoacoustic and Photothermal Phenomena 2009 (Leuven, Belgium).

Bogdanowicz Janusz, Dortu Fabian, Clarysse Trudo, Vandervorst Wilfried, Rosseel Erik, Nguyen Ngoc Duy, Shaughnessy Derrick, Salnik Alex and Nicolaides Lena

Nondestructive Extraction of Junction Depths of Active Doping Profiles from Photomodulated Optical Reflectance Offset Curves. Presented at the AVS Insight 2009 Workshop (Napa, California).

Bogdanowicz Janusz, Dortu Fabian, Clarysse Trudo, Vandervorst Wilfried, Shaughnessy Derrick, Salnik Alex, Nicolaides Lena and Opsal Jon

Impact of Inactive Dopants in Chemical Vapor Deposition Layers on Photomodulated Optical Reflectance. Presented at the EMRS 2008 Spring meeting (Strasburg, France).

Bogdanowicz Janusz, Dortu Fabian, Clarysse Trudo, Vandervorst Wilfried, Shaughnessy Derrick, Salnik Alex, Nicolaides Lena and Opsal Jon

Advances in Optical Carrier Profiling through High-Frequency Modulated Optical Reflectance. Presented at the AVS Insight 2007 Workshop (Napa, California).

Patents and Other Achievements

List of Patents and Patent Applications

Clarysse Trudo, Bogdanowicz Janusz

Method and Device to Quantify Active Carrier Profiles in Ultra-Shallow Semiconductor Structures,

Granting number : US7751035 (2010)

Published numbers : EP1922539 (2007), JP2009512180 (2009)

Bogdanowicz Janusz

Method For Determining the Doping Profile of a Partially Activated Doped Semiconductor Region,

Published numbers : US20100002236, EP2139033, JP2010034544 (2010)

Bogdanowicz Janusz

Method and Apparatus for Determining the Junction Depth of an Activated Doped Semiconductor Region Using Photomodulated Optical Reflectance Measurement Techniques

Application numbers : US12726173, EP10156708, JP2010061144 (2010)

Bogdanowicz Janusz, Clarysse Trudo, Vandervorst Wilfried *Method for Determining the Active Doping Concentration of a Doped Semiconductor Region*
Application numbers : EP10047, US61351705 (2010)

Bogdanowicz Janusz

Optical Measurement Method for Determining an Active Dopant Profile

Application number : US61366460 (2010)

Software

Dortu Fabian and Bogdanowicz Janusz

FSEM, a Finite element software for SEMiconductors, http://fsem.sourceforge.net/

Abbreviations

ac	Alternating Current
AFM	Atomic Force Microscopy
BF	Band Filling
BGN	Bandgap Narrowing
BX	BoxerCross
CMOS	Complementary Metal Oxide Semiconductor
CVD	Chemical Vapor Deposition
dc	Direct Current
FCA	Free Carrier Absorption
FPP	Four Point Probe
FQL	Flat Quasi Fermi Level
FSEM	Finite-element Solver for Semiconductors
FTIR	Fourier Transform Infrared Spectroscopy
IMEC	Interuniversitair Micro Electronica Centrum
IR	Infrared
ITRS	International Technology Roadmap for Semiconductors
JPV	Junction Photo Voltage
KK	Kramers Kronig
μ4PP	Micro Four Point Probe
MBIR	Model-Based Infrared Reflectometry
NIR	Near Infrared
PAI	Preamorphizing Implant
PMOR	Photomodulated Optical Reflectance
RsL	Sheet Resistance and Leakage
SIMS	Secondary Ion Mass Sprectroscopy
SRH	Shockley-Read-Hall
SRV	Surface Recombination Velocity
SSRM	Scanning Spreading Resistance Microscopy
TP	Therma-Probe
USJ	Ultra Shallow Junction

Symbols

k_b	Boltzmann Constant (J K^{-1})
q	Electron unit charge (C)
m	Electron unit mass (kg)
ϵ_0	Dielectric constant in vacuum (F m^{-1})
h	Planck constant (J s)
$\hbar$	$= h/(2\pi)$
c	speed of light in vacuum (m s^{-1})

Optics

$\tilde{\epsilon}$	Complex (relative) dielectric constant [1]
$\epsilon_1 = \Re(\tilde{\epsilon})$	Real part of the dielectric constant [1]
$\epsilon_2 = \Im(\tilde{\epsilon})$	Imaginary part of the dielectric constant [1]
$\tilde{n} = \sqrt{\tilde{\epsilon}}$	Complex refractive index [1]
$\Delta\tilde{n}$	Perturbation of the complex refractive index [1]
$n = \Re(\tilde{n})$	Real refractive index [1]
Δn	Perturbation of the real refractive index [1]
$k = \Im(\tilde{n})$	Extinction coefficient [1]
Δk	Perturbation of the extinction coefficient [1]
$\tilde{r}$	Reflection coefficient [1]
$\tilde{t}$	Transmission coefficient [1]
$R = \|\tilde{r}\|^2$	Reflectance [1]
ΔR	Perturbation of the reflectance [1]
R_{dc}	DC reflectance [1]
ΔR_{ac}	AC reflectance (a.u.)
λ_{laser}	Optical wavelength of the laser (laser = probe or pump) (m)
$\nu_{laser} = c/\lambda_{laser}$	Optical frequency of the laser (laser = probe or pump) (s^{-1})

$\omega_{laser} = 2\pi\nu_{laser}$	Optical angular frequency of the laser (laser = probe or pump) (s^{-1})
$\mathcal{R}$	Radius of the gaussian laser beam (m)
Π	Laser irradiance ($W\,m^{-2}$)
$\mathcal{P}$	Maximum in space of the laser irradiance (W m^{-2})
$\mathcal{P}^0$	Maximum in time of the laser irradiance (W m^{-2})
ω	Angular modulation frequency of the pump irradiance (s^{-1})
$\delta = \partial n/\partial T$	Thermorefractive coefficient (K^{-1})
$\beta = m\partial n/\partial N$	Electrorefractive Drude coefficient (kg m^3)

Materials

N	Free electron concentration (m^{-3})
N_{doping}	Free electron concentration due to doping (m^{-3})
ΔN	Excess free electron concentration due to optical injection (m^{-3})
P	Free hole concentration (m^{-3})
P_{doping}	Free hole concentration due to doping (m^{-3})
ΔP	Excess free hole concentration due to optical injection (m^{-3})
N_a^-	Ionized acceptor concentration (m^{-3})
N_d^+	Ionized donor concentration (m^{-3})
n_i	Intrinsic effective density (m^{-3})
T	Temperature of the lattice (K)
ΔT	Perturbation of the temperature of the lattice (K)
E_g	Bandgap energy (J)
E_c	Energy of the edge of the conduction band (J)
E_v	Energy of the edge of the valence band (J)
E_f	Quasi-Fermi energy level (J)
χ	Electron affinity (J)
ψ	Electrostatic potential (V)
$\vec{E}$	Electric field (V m^{-1})
m_e	Effective mass of electrons (kg)
m_h	Effective mass of holes (kg)
α_{BTB}	Band-to-band absorption coefficient (m^{-1})
G	Carrier generation rate ($m^{-3}s^{-1}$)
Rec	Carrier recombination rate ($m^{-3}s^{-1}$)
J	Current density (A cm^{-2})
σ	Electrical conductivity (ohm^{-1} m^{-1})
μ	Mobility ($m^2V^{-1}s^{-1}$)
D	Diffusivity (m^2s^{-1})
L	Decay length (m)
Λ	Wavelength of a damped wave (m)

ϕ	Phase of a damped wave (°)
R_s	Sheet resistance (Ohm/sq.)
X_j	Junction depth (m)

Sub- and superscripts

X_{sub} or X^{sub}	X in the substrate
X_l or X^l	X in the doped layer
X_0 or X^0	Constant mode of X
X_1 or X^1	Fundamental mode of X
X_n or X^n	Property X of electrons
X_p or X^p	Property X of holes
X_{pl} or X^{pl}	Property X of the carrier (plasma) distribution
X_{th} or X^{th}	Property X of the temperature distribution

Chapter 1
Introduction

At first sight, it may seem surprising that carrier profiling, i.e. *electrical* characterization, can be performed with an *optical* measurement technique. We will, however, show in this work that these two research fields are linked to each other. This chapter therefore briefly introduces both research domains. It is mostly a compilation of results and insights that have been gathered by colleagues and other researchers who I would like to deeply acknowledge, since their direct and indirect input has been essential to this work.

This Introduction starts with the motivation for this work in Sect. 1.1. The ultimate objective of this thesis being the development of an optical carrier-profiling tool, we then introduce in Sect. 1.2 the most commonly used techniques for the characterization of carrier and doping profiles in semiconductors. These techniques prove to be very useful for the final assessment of the developed technique. Further we shall see that each has its drawbacks and advantages. After discussing the destructive and/or contacting techniques in Sect. 1.2.1, focus is put upon non-destructive, non-contact techniques for carrier profiling in Sect. 1.2.2. Most important of all, Sect. 1.2.2.3 introduces the basics and state of the art of the Photomodulated Optical Reflectance (PMOR) technique. Finally, Sect. 1.3 describes the structure of this work.

1.1 Motivation

One of the many crucial issues in fabricating state-of-the-art Complementary Metal Oxide Semiconductor (CMOS) transistors is the precise electrical characterization of Ultra-Shallow Junctions (USJs) such as the source and drain extension regions (Fig. 1.1 (left)). The currently used low energy ion implantation and fast annealing techniques allow for very shallow (junction depth $X_j < 20$ nm) and highly actively doped (peak carrier concentration $N_{\text{peak}} > 10^{20}\,\text{cm}^{-3}$) regions [1] (Fig. 1.1 (right)). Furthermore, much effort is today placed in new techniques such as laser annealing to achieve higher concentration levels and steeper profiles. However, the shallower these

J. Bogdanowicz, *Photomodulated Optical Reflectance*, Springer Theses,
DOI: 10.1007/978-3-642-30108-7_1, © Springer-Verlag Berlin Heidelberg 2012

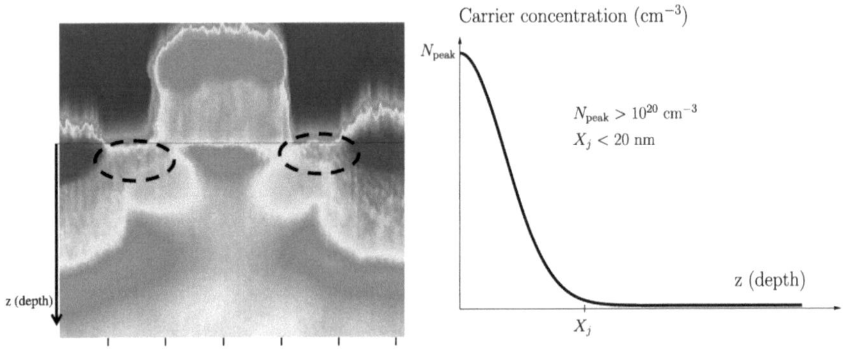

Fig. 1.1 *Left* Two-dimensional Scanning Spreading Resistance (Sect. 1.2.1.4) map of a planar MOS transistor. The circled regions are the source and drain extension regions, also called ultra-shallow junctions. *Right* Schematic of an ultra-shallow carrier profile such as typically found in the extension regions of modern transistors

doped regions become, the more critical their electrical characterization becomes. Small variations in e.g. the depth of these regions can indeed induce unacceptable changes in the performance of the final transistor [2]. The ability to monitor accurately the active doping profile of these USJs is therefore of critical importance.

Presently, dopant/carrier depth-profiling is usually based on a combination of Secondary Ion Mass spectrometry (SIMS; see Sect. 1.2.1.1) for the total profile (only chemical, not electrical information) and conventional Four-Point Probe (FPP; see Sect. 1.2.1.2) for sheet resistance, i.e. electrical activation. However, the combination of these techniques suffers from a significant number of disadvantages. First, SIMS is an off-line technique, applicable only on small pieces of material. Moreover, a measurement on one specific position on a wafer is rather time-consuming (dicing, sample preparation, measurement, data processing). Second, conventional FPP does not give any profile information and requires rather large analysis areas ($>1\,mm^2$). Besides, probe penetration leads to unreliable results on ultra-shallow profiles ($<30\,nm$) in the presence of well/halo implants.

Recently some new promising techniques have been emerging. First, Scanning Spreading Resistance Microscopy (SSRM; see Sect. 1.2.1.4) has shown very powerful two-dimensional carrier imaging capabilities. However, this off-line technique requires a complicated and critical sample preparation (destructive technique). Second, new FPP techniques have emerged, either with virtually zero penetration and a smaller measurement area [Micro Four-Point Probe (μFPP); see Sect. 1.2.1.5] or with no penetration whatsoever [Junction Photovoltage (JPV); see Sect. 1.2.2.1], but each of these also requires further research and development (sensitivity to surface oxide, surface states, reproducibility,...).

As a consequence, there is a real need for a new characterization technique for one-dimensional active dopant profiling which simultaneously:

- is sensitive to the active portion of doping profiles,

- can be used in-line,
- can perform the measurements quickly (5–10 min) and with a high reproducibility,
- is nondestructive and applicable to whole wafers (no sample preparation),
- has a high spatial resolution, i.e. only requires a minimal analysis area (measurement of small structures),
- extracts quickly the underlying profile with high precision and accuracy on both depth and concentration.

Among the plethora of existing measurement techniques, *optical reflection* techniques are the most promising to combine all the above mentioned requirements. On the one hand, *optical* probes do not require any physical contact with the sample, which makes them a very interesting candidate for in-line noninvasive measurements. On the other hand, for properly selected wavelengths, *reflection* techniques offer a very high depth resolution. We need to distinguish between *equilibrium* reflection techniques, such as Model-Based Infrared Reflectometry (MBIR; see Sect. 1.2.2.2) or Fourier-Transform Infrared Spectroscopy (FTIR; see Sect. 1.2.2.2), and *modulated* reflection techniques, such as Photomodulated Optical Reflectance (PMOR; see Sect. 1.2.2.3).

1.2 State of the Art of USJ Characterization

As highlighted in the motivation for this work, several techniques solve partially the problem which this work is focusing on. In this section, we therefore first briefly introduce these techniques. The description of each technique is strictly limited to their properties which are *relevant* in the framework of our investigations. More information can be found through the numerous added references. Further, only routinely used characterization techniques for silicon are discussed.

Since this work seeks a candidate for non-contact, non-destructive carrier profiling tool, it is logical to divide this section into two parts. Section 1.2.1 therefore reviews destructive and/or contact techniques. Section 1.2.2 then focuses on non-contact, non-destructive techniques. The latter section introduces PMOR, the main technique used in this work.

1.2.1 Contacting and/or Destructive Techniques

The techniques introduced in this section all either need to make a physical contact with the structure under investigation or even destroy it. This prohibits their use as fast nondestructive carrier profiling techniques. In other words, it is clear that none of these techniques can fulfill the six requirements mentioned in our motivation. They are only present in this work because they will help us assess and discuss the accuracy of the final developed technique.

Fig. 1.2 Principle of a SIMS measurement. The primary ion beam (*pink ions*) sputters the sample surface. The secondary ions, i.e. the ionized sputtered atoms (*green*, *blue* and *red*), are then accelerated towards a mass spectrometer for species determination

1.2.1.1 Secondary Ion Mass Spectrometry

Although Secondary Ion Mass Spectrometry (SIMS) does not give any activation information, it is often considered to be 'the' reference technique for dopant depth profiling. As shown schematically in Fig. 1.2, SIMS is based on the recording as a function of time of the ions sputtered from the sample by a primary ion beam (typically Cs^+ or O^- with typical energies 0.5–5 keV). The sputtered ions form a secondary ion beam which is accelerated towards a mass spectrometer for species determination, hence the name of the technique [3].

SIMS actually measures the number of *ionized sputtered atoms* of the considered species as a function of *time*. One therefore still needs to translate this raw information into a dopant profile, i.e. a *doping concentration* as a function of *depth*. Theoretically speaking, this can be achieved if one knows the ionization probability (i.e. the fraction of the sputtered atoms which have been ionized) and the sputter rate (i.e. the number of atoms sputtered by unit time) of the considered species in the considered matrix. In practice, the final crater depth is actually measured with a profilometer at the end of the SIMS measurement. The depth-scale of the doping profile is then obtained by linear interpolation, assuming a constant sputter rate. As for the doping concentration, it is obtained by comparison of the measured counts with calibration samples (same species, same matrix) [3].

Advantageously, this technique has a very high depth resolution. It, further, has a low detection limit, which allows for a very wide dynamic range. Up to four orders of magnitude of doping concentrations can typically be achieved on a doping profile, independently from the doping species or the matrix. However, five critical drawbacks of SIMS also need to be mentioned here. First, SIMS is a destructive technique (sputtering is destructive by essence). Second, due to ion mixing [4], SIMS always underestimates the profile slope (even though this problem is less at low energy [5]). Third, the presence of a surface oxide, which tends to initially accelerate sputtering [4], leads to a small underestimation (1–2 nm) of the profile depth, which is problematic especially for ultra-shallow junctions. Note that the impact of the last two observations leads to a difficult definition of the junction depth of very abrupt junctions. Fourth, SIMS gives only chemical information and no electrical information. Combining SIMS with FPP can, however, alleviate this limitation, as discussed in Sect. 1.2.1.3.

Finally, though not of application in this work which focuses on Si material, it also has to be mentioned that it is very complex to measure doping profiles in structures where the matrix changes (e.g. SiGe/Si), due to the very acute changes in sputter rate of the species of interest according to the matrix. In spite of all these drawbacks, the wide dynamic range of SIMS and its ability to measure with quite a good accuracy the local details of a dopant profile make this technique unavoidable.

1.2.1.2 Conventional Four-Point Probe

Conventional Four-Point Probe (FPP) is the reference technique for the measurement of the sheet resistance R_s of junction-isolated doped layers. By definition, this critical parameter reads

$$R_s = \frac{1}{\int_0^{X_j} q\mu N dz}, \tag{1.1}$$

where q is the electron unit charge, μ and N are respectively the mobility and concentration of the majority carriers, X_j is the thickness of the doped layer or (electrical) junction depth and z is depth. Formula (1.1) shows clearly that R_s is nothing but the integral of the carrier profile (convoluted with the mobility profile).

In-line FPP measurements are the most common resistance measurements. In these measurements, a current is injected between two probes and the induced potential difference between another two probes is measured (see Fig. 1.3b). Contrary to two-point probe measurements (Fig. 1.3a), the probes used to inject the current and those used to measure the potential difference are not the same so as to eliminate the impact of the probe and contact resistances on the measurements.

FPP is a very versatile [3] and fast technique (a few minutes for detailed wafer map). However, it suffers from a few drawbacks which are worth reminding. First and foremost, FPP only gives an integral of the profile of interest (zero-dimensional technique). This is the least detailed information which can be retrieved about a profile. Second, as illustrated in Fig. 1.3, the probes used in this technique apply a high load (5–100 g) on the sample, leading to some probe penetration into the sample (up to 30 nm for a 20 g load). In other words, as the junctions become shallower, FPP starts to measure not only the sheet resistance of the doped layer, but also the parallel substrate resistance [6] (substrate shorting). This leads to unacceptable underestimation of the sheet resistance. This effect is limited when the substrate doping is low enough ($<10^{16}$ cm^{-3}) [7].

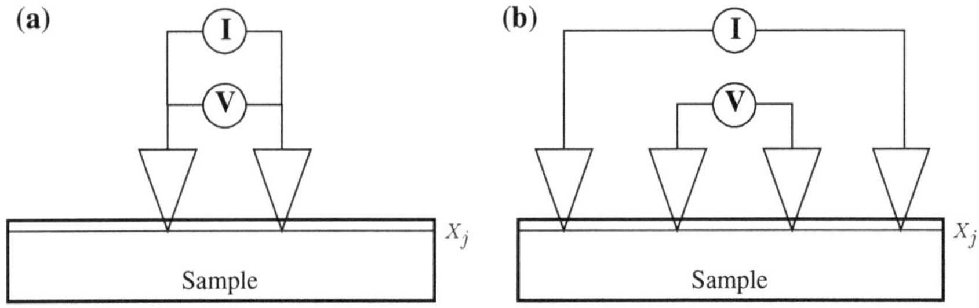

Fig. 1.3 **a** Schematic of a two-point probe resistance measurement set-up. The pins used to inject the current (I) are also used for the voltage (V) measurement. **b** Schematic of a four-point probe resistance measurement set-up. Different pins are used to inject the current and to measure the voltage

1.2.1.3 Combination of SIMS and FPP

As such, SIMS and FPP only offer partial information about the carrier profiles. SIMS measures the positions of the dopant atoms, whether they are activated or not and FPP measures the integral of the carrier profile. However, under the following conditions, full characterization of the carrier profile can be achieved by combining the two techniques. First, it must be assumed that the SIMS profile is only active up to a certain peak concentration $N_{\text{SIMS}}^{\text{peak}}$. The SIMS profile can therefore be cut off at this level $N_{\text{peak}}^{\text{SIMS}}$ above which all dopant atoms are assumed inactive (see Fig. 1.4). Second, the mobility of the majority carriers in the doped layer must be known. Ideally, it should be measured e.g. with a Hall technique [3, 8], as carried out in [9]. It can also be calculated based on the models available in the literature [10, 11]. The active doping profile can then be deduced by defining $N_{\text{SIMS}}^{\text{peak}}$ such that the integral of the SIMS profile below this level matches the sheet resistance measured with FPP. In other words, in order to derive the active SIMS profile, the following equation must be solved for $N_{\text{SIMS}}^{\text{peak}}$

$$\frac{1}{\underbrace{\int_0^{z_{\text{peak}}} q\mu N_{\text{SIMS}}^{\text{peak}} dz}_{q\mu z_{\text{peak}} N_{\text{SIMS}}^{\text{peak}}} + \int_{z_{\text{peak}}}^{X_j} q\mu N_{\text{SIMS}} dz} = R_s^{\text{FPP}}, \tag{1.2}$$

where R_s^{FPP} is the measured FPP sheet resistance and z_{peak} is the depth at which the measured SIMS profile reaches the concentration $N_{\text{SIMS}}^{\text{peak}}$. The resulting active SIMS profile of sample IIB-D07T5 is shown in Fig. 1.4, where the cut-off level has been determined using Klaassen's mobility model [10] assuming no carrier scattering on inactive dopants.

Two final remarks need to be made. First, it is of the utmost importance to realize that the position of $N_{\text{SIMS}}^{\text{peak}}$ strongly depends on the assumed mobility model and is

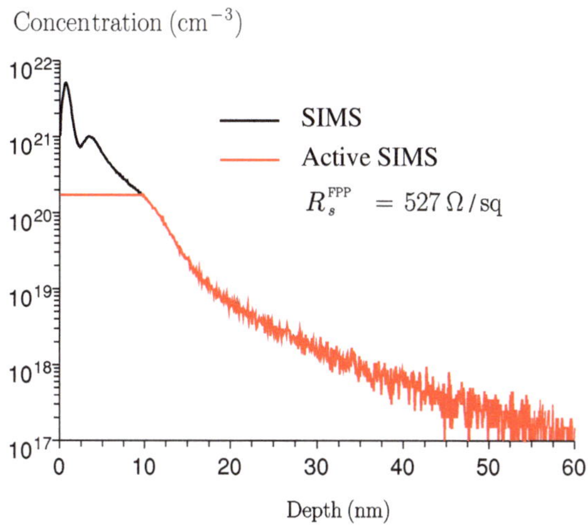

Fig. 1.4 Derivation from the measured SIMS profile (*black*) of sample IIB-D07T5 of the active SIMS profile (*red*), based on FPP measurement and Klaassen's mobility model assuming no scattering on inactive dopants

therefore not uniquely defined. We illustrate this limitation in Fig. 1.5a, where we compare the active SIMS profiles derived with two different assumptions on the mobility. Both models are based on Klaassen's formula for mobility [10]. First, like above, mobility model I assumes no scattering on inactive dopants. Even though this tends to overestimate mobility [9], it is the most widespread assumption. Mobility model II assumes that all active and inactive B atoms present in the SIMS profile act as scattering centers, which typically underestimates mobility.[1] As a result, the obtained N_{SIMS}^{peak} moves to higher values. Second, deriving the carrier profile actually still requires solving a one-dimensional Poisson equation. However, as shown in Fig. 1.5b, only very limited difference between both profiles is expected in the highly doped region (above $10^{18}\,cm^{-3}$). The differences are actually so minute that they are hindered by the noise on the active SIMS profile. For this reason, we will omit the difference between the carrier and active doping profiles in this thesis.

1.2.1.4 Scanning Spreading Resistance Microscopy

Scanning Spreading Resistance Microscopy (SSRM) is a very powerful resistivity-profiling technique. It has shown its capabilities on a wide variety of materials and structures [12–14]. SSRM is based on the measurement of the electrical resistance between a conductive Atomic Force Microscopy (AFM) tip, scanning the cross-section of the sample, and a backcontact (Fig. 1.6).

If the current path is carefully followed, the series of electrical resistances that are encountered is the following

[1] Calculated active doping concentrations much higher than the solubility limit can then be reached

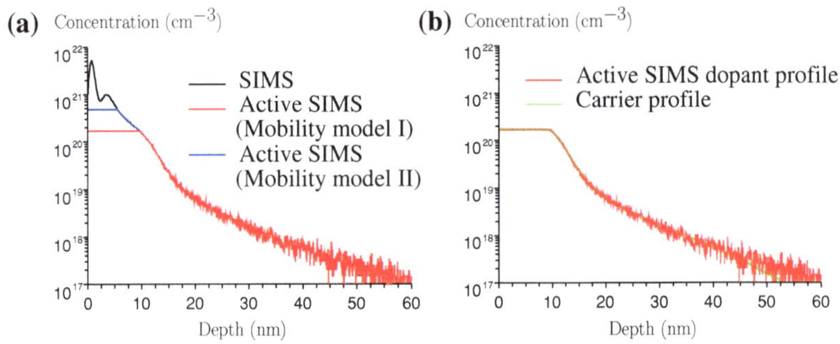

Fig. 1.5 **a** Comparison between the active SIMS profile of sample IIB-D07T5 assuming mobility model I, i.e. no carrier scattering on inactive dopants (*red*) and assuming mobility model II, i.e. all dopant atoms of the SIMS profile are scattering centers (*blue*). This shows that the calculated active dopant profile depends strongly on the used mobility model. **b** Comparison between the active SIMS profile (*red*) shown in Fig. 1.4 and the carrier profile derived by solving the one-dimensional Poisson equation. Negligible difference is observed in the high concentration region

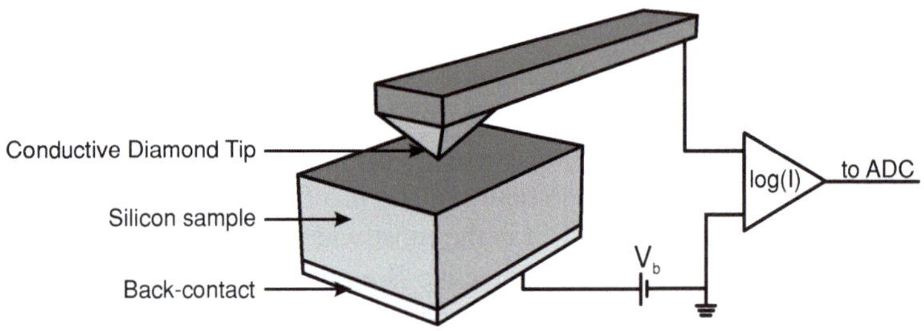

Fig. 1.6 Principle of an SSRM measurement. A voltage is applied between a very sharp conductive AFM tip and a backcontact. The measured current gives the spreading resistance related to the local carrier concentration via Eq. (1.4)

$$R_{\text{measured}} = R_{\text{tip}} + R_{\text{contact}} + R_{\text{spreading}} + R_{\text{sample}} + R_{\text{backcontact}} \tag{1.3}$$

where R_{tip} is the resistance of the cantilever and tip, R_{contact} is the contact resistance between the tip and the sample, $R_{\text{spreading}}$ is the resistance associated to the spreading of the current lines in the sample, R_{sample} is the sample resistance and $R_{\text{backcontact}}$ is the backcontact resistance. If one makes sure that the $R_{\text{spreading}}$ is dominant, the measured resistance reads [15]

$$R_{\text{measured}} = R_{\text{spreading}} = \frac{\rho}{4a} = \frac{1}{4aq(\mu_n N + \mu_p P)}, \tag{1.4}$$

where a is the radius of the electrical contact (typically 1 nm [16]), ρ is the local electrical resistivity of the sample. μ_n (resp. μ_p) and N (resp. P) are respectively the electron (resp. hole) mobility and concentration. Equation (1.4) shows that a two-dimensional resistance map obtained by scanning the conductive AFM tip over the sample cross section gives a direct image of the carrier distribution (Fig. 1.1(left)).

The simplicity of Eq. (1.4) is the strength of SSRM. SSRM being a direct measurement of the local carrier concentration, it suffers from fewer modeling artifacts than most techniques. This explains the success of SSRM when applied on new materials. With SSRM, you directly measure what you need. We will see that this is very different when it comes to optical techniques.

SSRM suffers, however, from some drawbacks. First of all, SSRM measurements are typically run on a (cleaved) cross section of the sample, which makes it a destructive and time-consuming technique. Second, SSRM is a relative technique relying on calibration samples for the quantification of the carrier concentrations, which renders the technique very dependent on the availability, quality and good characterization of these samples. Third, the Schottky nature of the contact and the presence of surface states [17] prove to have a dramatic impact on the measurement at low active doping concentration. Finally, it has to be mentioned that the tip resistance R_{tip} also limits the sensitivity of the technique in highly conductive regions. Except for the practical issues, SSRM also shows quantification problems both on the x-axis (depth scale) and on the y-axis (concentration scale) of the carrier profile. First, on the x-axis, the position of the oxide-Si interface (i.e. 'depth zero') is only determined by the strong decrease in measured resistance when moving from the oxide to doped Si. This can unfortunately not be done with a precision better than a few nanometers. For instance, the abruptness of the resistance profile of Fig. 1.7a, showing the resistance profile measured on sample IIB-D07T5, is about 1 nm/dec. The carrier profile of Fig. 1.7b has been obtained assuming the interface lies at a 12 nm depth. Second, on the y-axis, SSRM carrier extraction shows an uncertainty due to mobility. As highlighted in Eq. (1.4), the link between resistance and carrier concentration still requires the knowledge of mobility. Since SSRM does not measure mobility, it has to rely on existing mobility models [10, 11]. The profile shown in Fig. 1.7 assumes a carrier mobility of $100\,\mathrm{cm^2V^{-1}s^{-1}}$. Using a different mobility model, conclusions similar to those of Sect. 1.2.1.3 for the combination of SIMS and FPP would be drawn.

1.2.1.5 Micro Four-Point Probe

Micro Four-Point Probe (μFPP) is a recently developed sheet-resistance measurement technique based on FPP. Technically speaking, the improvement of μFPP over conventional FPP comes from the much smaller probe separation (down to 1.5 μm pitch vs. 1 mm in conventional FPP). We here only mention a few of the subsequent advantages, most of them being reviewed and detailed in Ref. [18]. First, obviously, thanks to its micron-size measurement head, μFPP is able to measure small structures (high spatial resolution). Second, due to the small pitch, the technique is much less sensitive to leakage [19] and consequently to the substrate resistance. Third, the low probe pressure renders μFPP a quasi-zero penetration tool. As a consequence, the problems due to the deep penetration of conventional probes are avoided. Finally, μFPP has shown very promising carrier-profiling capabilities on

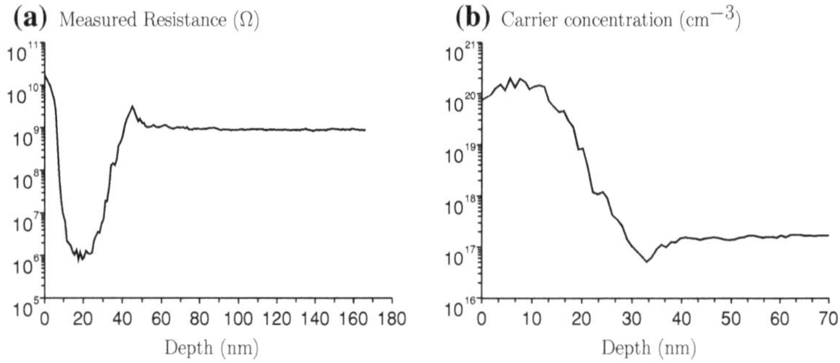

Fig. 1.7 **a** Raw data of the SSRM measurement of sample IIP-D07T5: resistance profile. **b** SSRM carrier profile of sample IIP-D07T5 assuming the oxide-Si interface is localized at a 12 nm depth and assuming a depth-independent mobility of $100\,\mathrm{cm^2\,V^{-1}\,s^{-1}}$

beveled structures [20]. However, this last application is still under intense investigation (reproducibility, starting point of the profile,...).

1.2.2 Non-Contacting and Non-Destructive Techniques

By essence, the contacting and/or destructive techniques introduced in the previous section cannot fulfill the six requirements for the ideal non-destructive carrier-profiling technique which is being sought. In this section, we propose to present the most recently developed techniques for the non-destructive electrical characterization of ultra-shallow junctions. In particular, we underline the drawbacks of each technique and show how this leads us to considering PMOR as the unique potential candidate to solve our needs.

1.2.2.1 Junction Photovoltage

Most of the sheet resistance values presented in this work have been obtained from Junction Photovoltage (JPV) measurements using, more particularly, the RsL embodiment [21, 22] of Frontier Semiconductor. Note that a similar tool has been developed by Semilab [23]. The principle of JPV measurements lies in the non-contact measurement of the position-dependent junction potential distribution resulting from the equilibrium between the generation of excess carriers injected by a low-irradiance light-emitting diode ($10^{-4}\,\mathrm{W\,cm^{-2}}$), which tend to increase the lateral electric field, and the drift of majority carriers in the highly doped region, which tends to decrease the lateral electric field. The drift current being inversely proportional to the sheet resistance, the measurement of the lateral distribution of JPV gives access to the sheet resistance [24] (Fig. 1.8).

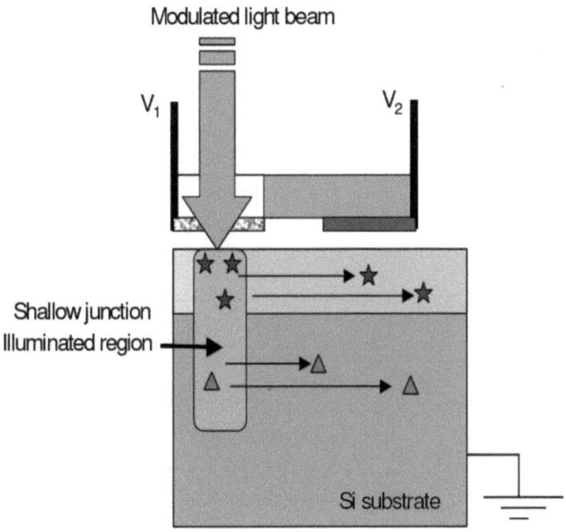

Fig. 1.8 Principle of a JPV measurement. The built-in potential is partially screened by excess carriers, as a result of the illumination of the sample. The majority carriers (*stars*) in the doped layer (*pink* region) consequently drift under the lateral electric field. This lateral drift current, which reduces the lateral electric field, is inversely proportional to the sheet resistance. Once the steady (-periodic) state is reached, the potential difference $V_1 - V_2$ is an image of the sheet resistance

This technique shows clear advantages. First, it is very fast (approximately 10 min for a 1,000 point map) and non-destructive. For moderate junction leakage ($\leq 10^{-3}$ Acm^{-2} [25]), the technique is able to separate the sheet resistance from the junction leakage[2] [26]. It is therefore an ideal technique for fast and accurate sheet resistance mapping of full wafers. We would, however, like to point out two critical drawbacks of this technique. First, just like all optical techniques, it is very sensitive to the properties of the measured material. Moving from Si to other materials (Ge,...) can therefore prove to be very complex (new models needed). It is frequently observed that, even in Si, USJs on a doped halo can lead to wrong or out-of-scale measurements [27]. It has been proposed that this is due to band-to-band tunneling. Second, the technique has a very poor spatial resolution due to its cm-sized probe head. As a consequence, JPV-based techniques, as currently implemented, will remain zero-dimensional electrical characterization technique and will not be able to measure ultra-shallow carrier profiles, even on a beveled surface. In other words, they cannot solve our need for a non-destructive carrier-profiling technique.

1.2.2.2 Model-Based Infrared Reflectometry and Fourier-Transform Infrared Ellipsometry

Reflectometry and ellipsometry are widely used techniques for the determination of the thickness and optical properties of thin films [28]. Both are based upon thin-film

[2] The name of the technique, "RsL", actually stands for Rs and Leakage measurement since it has been shown that the technique is able to measure leakage current simultaneously with sheet resistance. We will, however, not use or discuss that capability.

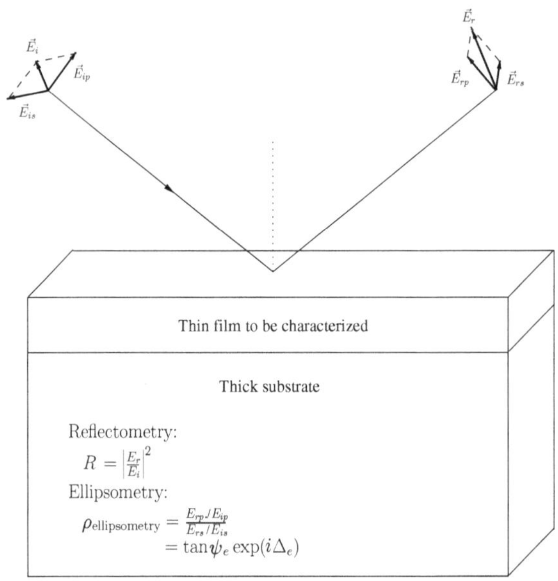

Fig. 1.9 Principle of measurement of thin-film reflectometry and ellipsometry. The reflected electric field $\vec{E}_r$ contains information about the thickness and optical properties of the thin film. Reflectometry measures the reflectance R while ellipsometry measures the state of polarization of the reflected light by means of the ellipsometric parameters, i.e. the amplitude tan ψ_e and phase Δ_e of $\rho_{\text{ellipsometry}}$. $\vec{E}_r$ (resp. $\vec{E}_i$), $\vec{E}_{rs}$ (resp. $\vec{E}_{is}$) and $\vec{E}_{rp}$ (resp. $\vec{E}_{ip}$) are respectively the reflected (resp. incident) electric field, the component of the reflected (resp. incident) field perpendicular to the plane of incidence and the component of the reflected (resp. incident) field parallel to the plane of incidence

interference, which is introduced in some more detail and studied in the framework of the perturbation theory in Chap. 2. Though both techniques are based upon the measurement of the light reflection on the sample under investigation, the measured parameters differ. While reflectometry measures the reflectance R of the sample, ellipsometry looks at the ellipsometric parameters, i.e. the state of polarization of the light reflected upon the sample.[3] as schematically shown in Fig. 1.9.

Model-Based Infrared Reflectometry (MBIR) and Fourier-Transform Infrared Ellipsometry (FTIR) are two specific implementations of the aforementioned techniques developed respectively by AMS [29] and SOPRA [30], wherein the wavenumber (i.e. the inverse of the wavelength) range of the incident light has been optimized to study doped silicon layers. One of the learnings derived from the Drude model (See Chap. 2) is that the impact of free carriers on the optical functions of Si is considerable starting from the infrared range down to smaller wavenumbers. Both companies

[3] The ellipsometric parameters are the amplitude tan ψ_e and phase Δ_e of the ratio of the parallel component E_{rp}/E_{ip} to the perpendicular component E_{rs}/E_{is} of the reflection coefficient. The parallel and perpendicular directions refer here to the plane of incidence

have therefore opted for wavenumbers included between 500 and 6,000 cm^{-1}, i.e. wavelengths included between roughly 1.5 and 20 μm.

As can be expected from the similarities between the two techniques, the results obtained by both techniques are comparable in many ways. The depth of deep (>100 nm) layers can be determined with quite a good agreement with SIMS [31]. The derived carrier concentrations, however, are always overestimated [32]. Further, shallow profiles cannot be measured due to the low sensitivity to junction depth (long wavelength). Repeatability measurements prove that these techniques are actually sensitive to the active dose [32]. Finally, both MBIR and FTIR use quite large (mm-size) beams and, as a consequence, have quite a low spatial resolution.

The drawbacks of these techniques clearly push towards PMOR and to the Therma-Probe (TP) embodiment more particularly. First, the reconstruction of ultra-shallow profiles, due to the low junction depth sensitivity, requires wavelengths shorter than those used in these infrared techniques. However, at shorter wavelengths, the impact of the presence of free carriers is strongly reduced (see e.g. the low numerical values for the variation in refractive index in Chap. 3), which in turn reduces the signal-to-noise ratio. TP solves this problem by using a high-power light source so as to keep a sufficient signal-to-noise ratio and, consequently, high enough measurement reproducibility. In other words, TP being a high-irradiance reflectometry measurement, it partially solves the problems encountered by FTIR and MBIR. Second, PMOR measurements offer intrinsically an independent junction depth measurement. As we will show in Chap. 7, the AC reflectance signal is indeed very sensitive to the depth of the carrier profile. Further, no direct correlation with dose can be observed. The AC reflectance could therefore be used as the independent depth measurement that FTIR and MBIR seem to lack. Note, finally, that the shorter wavelength also enhances the spatial resolution since a smaller beam radius is then allowed (Rayleigh's diffraction limit [33]).

As a conclusion, it seems that PMOR has more to offer than MBIR or FTIR. This would make PMOR an ideal and unique candidate possibly fulfilling the six criteria mentioned in the motivation for this work.

1.2.2.3 Photomodulated Optical Reflectance

Photomodulated Optical Reflectance (PMOR) is a pump-probe technique wherein a probe laser measures not only the sample reflectance (i.e. like reflectometry), which we call the DC reflectance R_{dc}, but also the variations in reflectance induced by a modulated pump laser, the so-called modulated or AC reflectance ΔR_{ac}. The modulated reflectance signal is the most commonly used signal, which is why the technique bears its name. Figure 1.10 schematically describes the principle of the technique.

Different companies (KLA-Tencor, Semilab, Xitronix) and research groups (e.g. Center for Advanced Diffusion-Wave Technologies of the University of Toronto and Laboratoire d'Instrumentation of the Universite Pierre et Marie Curie in Paris) have implemented the PMOR technique either in fab-ready or test-bench tools. Though PMOR initially showed capabilities for the determination of various transport

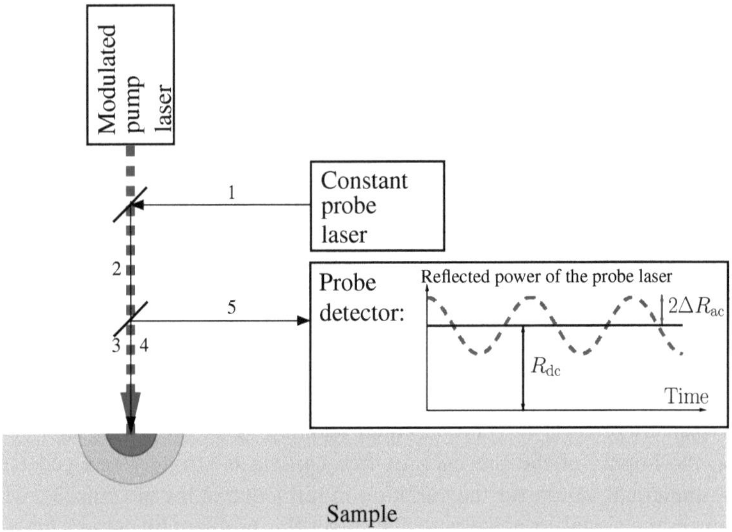

Fig. 1.10 Principle of Photomodulated Optical Reflectance (PMOR). A modulated pump laser shines on the sample to modify the local refractive index. A constant probe laser measures the sample reflectance, i.e. the so-called DC reflectance R_{dc}, simultaneously with the variations in reflectance induced by the modulated pump laser, i.e. the so-called modulated or AC reflectance ΔR_{ac}. The numbers indicate the path followed by the probe laser beam

parameters of bulk and thin-film metals, semiconductors or superconductors [34–38] (e.g. thermal conductivity, carrier diffusivity, carrier lifetime), commercial implementations mostly focus on statistical process control (SPC) monitoring of ion implantation [39–47]. Importantly, intensive research has also been carried out to improve the understanding of the complex behavior of the AC reflectance on bulk semiconductors [48–56].

Though there has been some discussion about the original paternity of the idea, as far as the literature can testify, interest in the use of PMOR for carrier profiling initially started with the work of Borden [57–59] and Nicolaides [39, 40] respectively with the BX10 tool of Boxercross (now owned by Semilab) and with the TP630XP tool of Thermawave (now owned by KLA-Tencor). The conclusions of these investigations already proved experimentally the great promises of PMOR for non-destructive carrier profiling. Sub-nanometer junction depth resolution seemed finally to be in the reach of optical techniques. However, theoretically speaking, there was only qualitative understanding of the dominant physical processes occurring during a PMOR measurement on a non-homogeneously doped silicon sample. Soon imec became involved [47, 60–63] and the combined theoretical and experimental work of Dortu [64–68] on the BX10 tool got the theory ever closer to experimental data. However, when it came to retrieving information about carrier profiles, the BX10 tool seemed to lack information. The ability of this tool to measure power curves, i.e. vary its pump power and measure the induced variations in ΔR_{ac}, did not seem to help much. Besides, as time passed by, the probe wavelength of the BX10 (980 nm) was becoming obsolete compared to the junction depths required in real devices. In the meantime, the higher modulation frequency (1 MHz vs. 2 kHz) of the TP630XP had appeared as a potential solution to the problem by offering two signals (amplitude and phase of ΔR_{ac}) rather than one [7]. Its shorter probe wavelength (670 nm) also improved significantly the sensitivity to sub-30 nm junction depths.

These two observations, together with the high repeatability of the measurement (Appendix A.2), were the motivations which initially triggered this work. The 'two-signal' approach looked promising though, as we shall prove, the phase actually does not contain extra information about the profiles (plasma wave has a unique phase and wavelength; see e.g. Sect. 6.2). Nevertheless, we will show that TP630XP offers additional information about the profiles. First, it has the capability to measure offset curves, i.e. separate the pump and probe beams and measure the variations in ΔR_{ac} (Chap. 6). Besides offering a much better visibility to the processes involved in the carrier and heat transport (e.g. you can *measure* the ambipolar diffusivity, you do not need to *model* it, see Sect. 6.1), offset curves also help decouple the sensitivity to carrier concentration and to junction depth (see Sect. 7.1.1). Finally, R_{dc} being calibrated in TP630XP, its direct use is facilitated. As we shall see, this last improvement was also a determining factor (see Sect. 7.2).

Though we will come back to these achievements in the course of this thesis as well as in the Conclusions, some can already be mentioned. First, on the experimental side, the work included in this thesis has contributed to making PMOR a reproducible and quantitative technique (see Appendix A.2). Further, the developed procedure for the signal calibration on undoped samples (Sects. 6.1, 7.2) has broadened the applicability window of the technique to implanted profiles with and without preamorphization. Before our investigations, indeed, most work had been dedicated to box-like profiles. Second, on the theoretical side, improved understanding also comes out of this work thanks to the derivation of simplified transport equations to describe the general ambipolar transport of carriers (Sect. 4.1.1) and heat (Sect. 4.1.2) in semiconductors, with a thorough discussion of their time-dependent effects (Sect. 4.1.3). Similarly, the compact models of Chap. 6 give direct understanding of the general behavior of ΔR_{ac} and R_{dc}.

As mentioned above, this work focuses on the TP630XP tool. In other words, while we often refer to the technique as PMOR, all shown experimental data have been measured with TP630XP. In particular, the TP630XP (or simply TP) tool is a PMOR implementation with:

- a 670 nm (1.85 eV) probe laser with 2.5 mW power focused onto an 0.5 μm beam radius on the sample
- a 790 nm (1.57 eV) pump laser with 13.5 mW power modulated with a 1 MHz modulation frequency and focused onto an 0.5 μm beam radius on the sample
- the ability to separate the laser beam with a maximum of 4 μm separation on the sample.

More detailed examples of the experimental setup of PMOR tools can be found in Ref. [69, 70]. The actual setup of the TP630XP being confidential, we are not allowed to disclose it in this work.

1.3 Outline

The ultimate objective of this work, besides the fundamental investigation of PMOR, is to develop a quantitative carrier profiling technique out of the TP tool such as it is described above in Sect. 1.2.2.3 without any hardware change. For this purpose, we need to understand why, how and also how much a pump laser can modify the reflectance of a silicon sample. We therefore propose to build up the necessary understanding of the processes involved during a TP measurement in the following sequence.

In Chap. 2, we look at the fundamental process accounting for a variation in the reflectance of an arbitrary sample. In particular, looking first qualitatively at the relationship between refractive index and reflectance, we sequentially derive analytical expressions for the variation in reflectance assuming a refractive index perturbation with various shapes. One important message of this chapter is that a *perturbation of the reflectance* is always a consequence of a *perturbation of the refractive index*.

In Chap. 3, we investigate the possible causes for a variation in the refractive index of silicon. This chapter shows that a *perturbation in refractive index* is to be expected as a result of the *presence of free carriers or the change in temperature*. In particular, we quantify these variations for the specific case of the TP probe laser.

In Chap.4, we look at how light and, more specifically, a high-irradiance *modulated pump laser* and a high-irradiance *constant probe laser* generate carriers and heat in homogeneously doped silicon samples. We derive and solve the coupled transport equations for these two quantities in order to quantitatively determine the time-dependent and position-dependent *excess carrier and temperature distributions* induced by the lasers.

In Chap. 5, we extend the transport problem to the case of non-homogeneously doped silicon sample (with particular emphasis on samples presenting highly doped ultra-shallow layers at their surface).

In Chap. 6, we review the conclusions of Chaps. 2, 3, 4 and 5 in order to derive a complete *physical model* explaining the behavior of the two measured signals in PMOR, i.e. R_{dc} and ΔR_{ac}. The model is assessed based on the comparison with TP signals measured on homogeneously doped silicon substrates with various doping concentrations as well as doped layers with varying doping concentration and junction depth.

Based on the learnings of all the previous chapters, Chap. 7 develops and studies methods to derive *information about carrier profiles* from PMOR measurements.

References

1. International technology roadmap for semiconductors. International Technology Roadmap for Semiconductors, http://www.itrs.net.http://www.itrs.net/reports.html
2. S. Severi, Ultra Shallow Junctions Formation for the Ultimate Scaling Limit: Physics, Fabrication and Characterization. PhD thesis, Katholieke Universiteit Leuven, 2006
3. D.K. Schroder, *Semiconductor Material and Device Characterization, Chap. 9* (Wiley-IEEE Press, New York, 2006)
4. W. Vandervorst, T. Janssens, B. Brijs, T. Conard, C. Huyghebaert, J. Fruhauf, A. Bergmaier, G. Dollinger, T. Buyuklimanli, J.A. VandenBerg, K. Kimura, Errors in near-surface and interfacial profiling of boron and arsenic. Appl. Surf. Sci. **231–232**, 618–631 (2004)
5. W. Vandervorst, Semiconductor profiling with sub-nm resolution: challenges and solutions. Appl. Surf. Sci. **255**(4), 805–812 (2008)
6. T. Clarysse, F. Dortu, D. Vanhaeren, I. Hoflijk, L. Geenen, T. Janssens, R. Loo, W. Vandervorst, B.J. Pawlak, V. Ouzeaud, C. Defranoux, V.N. Faifer, M.I. Current, Accurate electrical activation characterization of cmos ultra-shallow profiles. Mater. Sci. Eng. B-Solid State Mater. Adv. Technol. **114–115**, 166–173 (2004)
7. T. Clarysse, W. Vandervorst, M. Bakshi, L. Nicolaides, A. Salnik, J. Opsal, Towards nondestructive carrier depth profiling. J. Vac. Sci. Technol. B **24**(3), 1139–1146 (2006)
8. D.H. Petersen, O. Hansen, R. Lin, P.F. Nielsen, Micro-four-point probe hall effect measurement method. J. Appl. Phys. **104**(1), 013710-1-10 (2008)
9. T. Clarysse, J. Bogdanowicz, J. Goossens, A. Moussa, E. Rosseel, W. Vandervorst, D.H. Petersen, R. Lin, P.F. Nielsen, O. Hansen, G. Merklin, N.S. Nennett, N.E.B. Cowern, On the analysis of the activation mechanisms of sub-melt laser anneals. Mater. Sci. Eng. B **154–155**, 24–30 (2008)
10. D.B.M. Klaassen, A unified mobility model for device simulation. 1. model-equations and concentration-dependence. Solid-State Electron. **35**(7), 953–959 (1992)
11. American Society for Testing and Materials, *Annual Book of ASTM Standards* (American Society for Testing and Materials, Philadelphia, 1987)
12. P. Eyben, Scanning Spreading Resistance Micorscopy: High resolution two-dimensional carrier profiling of semiconductor structures. Ph.D. thesis, KUleuven, 2004
13. J. Mody, R. Duffy, P. Eyben, J. Goossens, A. Moussa, W. Polspoel, B. Berghmans, M.J.H. van Dal, B.J. Pawlak, M. Kaiser, R.G.R. Weemaes, W. Vandervorst, Experimental studies of dose retention and activation in fin field-effect-transistor-based structures (vol 28, pg c1h5, 2010). J. Vac. Sci. Technol. B **28**(3), 648–648 (2010)
14. T. Clarysse, G. Brammertz, D. Vanhaeren, P. Eyben, J. Goossens, F. Clemente, M. Meuris, W. Vandervorst, R. Srnanek, R. Kinder, B. Sciana, D. Radziewicz, Z.Q. Li, Accurate carrier profiling of n-type gaas junctions. Mater. Sci. Semicond. Process. **11**(5–6), 259–266 (2008)
15. P. De Wolf, T. Clarysse, W. Vandervorst, Quantification of nanospreading resistance profiling data, vol. 16, pp. 320–326. AVS, 1998, http://link.aip.org/link/?JVB/16/320/1
16. D. Alvarez, J. Hartwich, M. Fouchier, P. Eyben, W. Vandervorst, Sub-5-nm-spatial resolution in scanning spreading resistance microscopy using full-diamond tips. Appl. Phys. Lett. **82**(11), 1724–1726 (2003)
17. P. Eyben, S. Denis, T. Clarysse, W. Vandervorst, Progress towards a physical contact model for scanning spreading resistance microscopy. Mater. Sci. Eng. B-Solid State Mater. Adv. Technol. **102**(1–3), 132–137 (2003)
18. D.H. Petersen, O. Hansen, T.M. Hansen, P. Boggild, R. Lin, D. Kjaer, P.F. Nielsen, T. Clarysse, W. Vandervorst, E. Rosseel, N.S. Bennett, N.E.B. Cowern, Review of electrical characterization of ultra-shallow junctions with micro four-point probes. J. Vac. Sci. Technol. B **28**(1), C1C27–C1C33 (2010)
19. C.L. Petersen, R. Lin, D.H. Petersen, P.F. Nielsen, Micro-scale sheet resistance measurements on ultra shallow junctions. *2006 14th International Conference on Advanced Thermal Processing of Semiconductors (IEEE Cat No. 06EX1530)*, (2006) pp. 153–158

20. T. Clarysse, P. Eyben, B. Parmentier, B. Van Daele, A. Satta, W. Vandervorst, R. Lin, D.H. Petersen, P.F. Nielsen, Advanced carrier depth profiling on si and ge with micro four-point probe. J. Vac. Sci. Technol. B **26**(1), 317–321 (2008)
21. V.N. Faifer, M.I. Current, T.M.H. Wong, V.V. Souchkov, Noncontact sheet resistance and leakage current mapping for ultra-shallow junctions. J. Vac. Sci. Technol. B: Microelectron. Nanom. Struct. **24**(1), 414–420 (2006). http://link.aip.org/link/?JVB/24/414/1
22. T.M.H. Wong T. Nguyen M.I. Current, V.N. Faifer, A. Koo. Non-contact sheet resistance and leakage current monitoring of multi-implant, ultra-shallow junctions: Doping and damage effects for ms-anneals
23. F. Korsos, K. Kis-Szabo, E. Don, A. Pap, T. Pavelka, C. Laviron, M. Pfeffer, Junction photovoltage metrology and high resolution mapping of ion implants electrically isolated from the wafer surface. Ion Implant. Technol. **2008**(1066), 113–116 (2008)
24. V.N. Faifer, M.I. Current, D.K. Schroder, Characterization of ultrashallow junctions using frequency-dependent junction photovoltage and its lateral attenuation. Appl. Phys. Lett. **89**(15), 151123 (2006)
25. Frontier Semiconductor. Operational Manual: RsL 100
26. V.N. Faifer, M.L. Current, T. Nguyen, T.M.H. Wong, V.V. Souchkov, Non-contact measurement of sheet resistance and leakage current: applications for usj-sde/halo junctions. Extended Abstracts of the Fifth International Workshop on Junction Technology (IEEE Cat. No.05EX1126), 2006, pp. 45–48
27. T. Clarysse, A. Moussa, B. Parmentier, J. Bogdanowicz, W. Vandervorst, H. Bender, M. Pfeffer, M. Schellenberger, P.F. Nielsen, S. Thorsteinsson, R. Lin, D. Petersen, Photovoltage versus microprobe sheet resistance measurements on ultrashallow structures. J. Vac. Sci. Technol. B **28**(1), C1C8–C1C14 (2010)
28. H. Fujiwara, *Spectroscopic Ellipsometry: Principles and Applications* (Wiley, Chichester, 2007)
29. M. Gostein, P.A. Rosenthal, A. Maznev, A. Kasic, P. Weidner, P.Y. Guittet, Measuring deep-trench structures with model-based ir. Solid State Technol. **49**(3), 38 (2006)
30. C. Defranoux, T. Emeraud, S. Bourtault, J. Venturini, P. Boher, M. Hernandez, C. Laviron, I. Noguchi, Infrared spectroscopic ellipsometry applied to the characterization of ultra shallow junction on silicon and soi. Thin Solid Films **455–456**, 150–156 (2004)
31. B. Senitzky, S.P. Weeks, Infrared reflectance spectra of thin epitaxial silicon layers. J. Appl. Phys. **52**(8), 5308–5314 (1981), http://link.aip.org/link/?JAP/52/5308/1
32. A. Maznev, Measuring usj samples with model-based infrared spectroscopic reflectometry (mbir). Technical report, AMS, 2006
33. M. Born, E. Wolf, *Principles of Optics* (Pergamon, Oxford, 1970)
34. I. Barbereau, B.C. Forget, D. Fournier, Characterization of electronic transport properties in semiconductors by scanning photothermal microscopy. Prog. Nat. Sci. **6**, S479–S482 (1996)
35. B.C. Forget, D. Fournier, Electronic transport-properties characterization of silicon-wafers by modulated photoreflectance. Journal de Physique IV **1**(C6), 277–282 (1991)
36. B.C. Forget, I. Barbereau, D. Fournier, S. Tuli, A.B. Battacharyya, Electronic diffusivity measurement in silicon by photothermal microscopy. Appl. Phys. Lett. **69**(8), 1107–1109 (1996)
37. D. Fournier, A.C. Boccara, Thermal wave probing of the optical electronic and thermal-properties of semiconductors. Mater. Sci. Eng. B-Solid State Mater. Adv. technol. **5**(2), 83–88 (1990)
38. B. Li, L. Pottier, J.P. Roger, D. Fournier, E. Welsch, Thermal characterization of film-on-substrate systems with modulated thermoreflectance microscopy. Rev. Sci. Instrum. **71**(5), 2154–60 (2000)
39. L. Nicolaides, A. Salnick, J. Opsal, Study of low energy implants for ultrashallow junctions using thermal wave and optical techniques. Rev. Sci. Instrum. **74**(1), 563–565 (2003)
40. L. Nicolaides, A. Salnick, J. Opsal, Nondestructive analysis of ultrashallow junctions using thermal wave technology. Rev. Sci. Instrm. **74**(1), 586–588 (2003)
41. A. Salnick, J. Opsal, Quantitative photothermal characterization of ion-implanted layers in si. J. Appl. Phys. **91**(5), 2874–2882 (2002)

42. G. Smets, E. Rosseel, G. Sterckx, J. Bogdanowicz, W. Vandervorst, D. Shaughnessy, Transfer from rs-based to pmor-based ion implantation process monitoring. AIP Conf. Proc. **1321**, 426–31 (2011)
43. P. Borden, C. Ferguson, D. Sing, L. Larson, L. Bechtler, K. Jones, P Gable, In-line characterization of preamorphous implants (pai). Technical report, BoxerCross, 2000
44. C. Christofides, I.A. Vitkin, A. Mandelis, Photothermal reflectance investigation of processed silicon. i. room-temperature study of the induced damage and of the annealing kinetics of defects in ion-implanted wafers. J. Appl. Phys. **67**(6), 2815–21 (1990)
45. B.C. Forget, D. Fournier, Characterization of implanted silicon-wafers by the nonlinear photoreflectance technique. Mater. Sci. Eng. B-Solid State Mater. Adv. Technol. **24**(1–3), 199–202 (1994)
46. B.C. Forget, D. Fournier, V.E. Gusev, Non-linear recombination processesapplication to quantitative implantation characterization. J. de Physique **4**(C7), 155–158 (1994)
47. W. Vandervorst, T. Clarysse, B. Brijs, R. Loo, Y. Peytier, B.J. Pawlak, E. Budiarto, and P. Borden, Carrier illumination as a tool to probe implant dose and electrical activation. in *2003 International Conference on Characterization and Metrology for ULSI Technology*, ed. by D.G. Seiler, A.C. Diebold, T.J. Shaffner, R. McDonald, S. Zollner, R.P. Khosla, E.M. Secula, vol. 683 (AIP, 2003) pp. 758–763. http://link.aip.org/link/?APC/683/758/1
48. A. Rosencwaig, J. Opsal, W.L. Smith, D.L. Willenborg, Detection of thermal waves through optical reflectance. Appl. Phys. Lett. **46**(11), 1013–1015 (1985)
49. D. Fournier, C. Boccara, A. Skumanich, N.M. Amer, Photothermal investigation of transport in semiconductorstheory and experiment. J. Appl. Phys. **59**(3), 787–795 (1986)
50. J. Opsal, Thermal and plasma waves in semiconductors. Rev. Progr. Quant. Nondestruct. Eval. **6B**, 1339–1346 (1987)
51. J. Opsal, M.W. Taylor, W.L. Smith, A. Rosencwaig, Temporal behavior of modulated optical reflectance in silicon. J. Appl. Phys. **61**(1), 240–248 (1987)
52. J. Opsal, A. Rosencwaig, Theory of bulk and near surface effects on the modulated optical reflectance in silicon. Photoacoustic and Photothermal Phenomena **58**, 224–228 (1988)
53. R.E. Wagner, A. Mandelis, A generalized calculation of the temperature and drude photomodulated optical reflectance coefficients in semiconductors. J. Phys. Chem. Solids **52**(9), 1061–1070 (1991)
54. R.E. Wagner, A. Mandelis, Quantitative photomodulated thermoreflectance studies of germanium and silicon semiconductors. J. de Physique IV **4**(C7), 141–144 (1994)
55. A. Mandelis, R.E. Wagner, Quantitative deconvolution of photomodulated thermoreflectance signals from si and ge semiconducting samples. Jpn. J. Appl. Phys. Part 1 **35**(3), 1786–1797 (1996)
56. C. Christofides, A. Othonos, E. Loizidou, Photomodulated thermoreflectance investigation at elevated temperatures: plasma versus thermal effect. Appl. Phys. Lett. **82**(7), 1132–4 (2003)
57. US Patent 6323951 B1 (Nov. 2001): Apparatus and method for determining the active dopant profile in a semiconductor wafer, 2001
58. P. Borden, R. Nijmeijer, J. P. Li, L. Bechtler, K. Lingel, Nondestructive profile measurements of annealed shallow implants. J. Vac. Sci. Technol. B: Microelectron. Nanom. Struct. textbf18(1), 602–604, (2000) http://link.aip.org/link/?JVB/18/602/1
59. P. Borden, Optical, non-destructive characterization of ultra-shallow junctions. Mater. Sci. Semiconduct. process. **4**(1–3), 9–14 (2001)
60. T. Clarysse, W Vandervorst, R. Lindsay, P. Borden, E. Budiarto, J. Madsen, N.J. Nijmeijer Junction and profile analysis using carrier illumination. Technical report, IMEC, 2003
61. T. Clarysse, R. Lindsay, W. Vandervorst, E. Budiarto, P. Borden. Carrier illumination for characterization of ultrashallow doping profiles. J. Vac. Sci. Technol. B: Microelectron. Nanom. Struct. **22**(1), 439–443 (2004) http://link.aip.org/link/?JVB/ww/439/1
62. T. Clarysse, F. Dortu, W. Vandervorst, Patent: Method for the independent extraction of the carrier concentration level and electrical junction depth in carrier illumination. Technical report, IMEC, 2004

63. W. Vandervorst, T. Clarysse, N. Duhayon, P. Heyben, T. Hantschel, M. Xu, T. Janssens, Ultra shallow junction profiling. in *IEDM*, (2000) pp. 429–432
64. F. Dortu, T. Clarysse, R. Loo, B. Pawlak, R. Delhougne, W. Vandervorst, Progress in the physical modeling of carrier illumination. J. Vac. Sci. Technol. B **24**(3), 1131–1138 (2006)
65. F. Dortu, T. Clarysse, R. Loo, W. Vandervorst, Extracting active dopant profile information from carrier illumination power curves. J. Vac. Sci. Technol. B **24**(1), 375–380 (2006)
66. F. Dortu, J. Bogdanowicz, T. Clarysse, W. Vandervorst, Nonlinear study of photoelectrothermal modulated optical reflectance for active dopant profile extraction. J. Appl. Phys. **101**, 053107 (2007)
67. F. Dortu, J. Bogdanowicz, T. Clarysse, Impact of band gap narrowing and surface recombination on photoelectrothermal modulated optical reflectance power curves. J. Vac. Sci. Technol. B **26**, 322–332 (2008)
68. F. Dortu, Low-Frequency Modulated Optical Reflectance for the One-Dimensional Characterization of Ultra-Shallow Junctions. Ph.D. thesis, Katholieke Universiteit Leuven, 2009
69. L. Nicolaides, A. Salnik, J. Opsal, Position modulated optical reflectance measurement system for semiconductor metrology. (7212288), May 2007, http://www.freepatentsonline.com/7212288.html
70. C.G. Wells , A. Rosencwaig, W.L .Smith, Apparatus for evaluating thermal and electrical characteristics in a sample. (5228776), July 1993, http://www.freepatentsonline.com/5228776.html

Chapter 2
Theory of Perturbation of the Reflectance

In this chapter, we assume a small perturbation of the complex refractive index $\tilde{n}(x, y, z, t)$ is somehow generated (e.g. by doping, optically injected carriers and/or heat,...) in a homogeneous silicon sample and we calculate how this perturbation impacts the reflectance $R(x, y, t)$ of a probe laser shining on the sample.

In all generality, the perturbed refractive index reads

$$\tilde{n}(x, y, z, t) = \tilde{n}_0 + \Delta\tilde{n}_0(x, y, z) + \underbrace{\Delta\tilde{n}_1(x, y, z)\cos[\omega t - \phi_1(x, y, z)]}_{\text{fundamental}} + \sum_{j=2}^{+\infty} \underbrace{\Delta\tilde{n}_j(x, y, z)\cos[j\omega t - \phi_j(x, y, z)]}_{j\text{th harmonic}}, \tag{2.1}$$

where $\tilde{n}_0$ is the refractive index of the unperturbed silicon sample, $\Delta\tilde{n}_0(x, y, z)$ is the time-independent component of the perturbation of the refractive index, $\Delta\tilde{n}_1(x, y, z)$ and $\phi_1(x, y, z)$ are respectively the amplitude and phase of the fundamental mode of the perturbation, $\Delta\tilde{n}_j(x, y, z)$ and $\phi_j(x, y, z)$ $(j = 2, 3, \ldots)$ are respectively the amplitude and phase of the jth harmonic of the perturbation. Note that $\tilde{n}_0$ and $\Delta\tilde{n}_j(x, y, z)$ $(j = 0, 1, \ldots)$ are all complex numbers. In this chapter and the following, ω can be understood in all generality as the fundamental angular frequency of the refractive index perturbation but, it will become clear in Chap. 4 that ω is the angular modulation frequency of the pump laser irradiance.

As a result of this perturbation, a probe laser shining on the sample is reflected with a perturbed reflectance $R(x, y, t)$ such that

$$R(x, y, t) = R_0 + \Delta R_0(x, y) + \underbrace{\Delta R_1(x, y)\cos[\omega t - \Phi_1(x, y)]}_{\text{fundamental}} + \sum_{j=2}^{+\infty} \underbrace{\Delta R_j(x, y)\cos[j\omega t - \Phi_j(x, y)]}_{j\text{th harmonic}}, \tag{2.2}$$

J. Bogdanowicz, *Photomodulated Optical Reflectance*, Springer Theses,
DOI: 10.1007/978-3-642-30108-7_2, © Springer-Verlag Berlin Heidelberg 2012

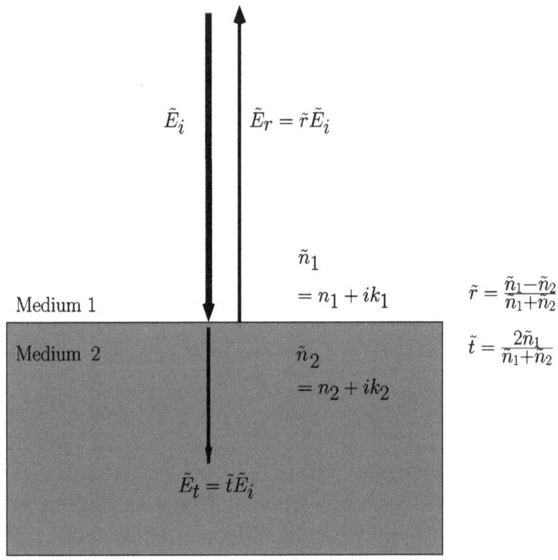

Fig. 2.1 An electromagnetic wave with electric field $\tilde{E}_i$ incident on an interface between two media of respective refractive indices $\tilde{n}_1$ and $\tilde{n}_2$ is partially reflected with electric field $\tilde{E}_r$ and transmitted with electric field $\tilde{E}_t$. The ratio between the reflected (resp. transmitted) and incident electric fields is given by $\tilde{r}$ (resp. $\tilde{t}$) which follows Fresnel's reflection formula (2.3) [resp. (2.4)]

where R_0 is the reflectance of the unperturbed Si sample, $\Delta R_0(x, y)$ is the time-independent component of the perturbation of the reflectance, $\Delta R_1(x, y)$ and $\Phi_1(x, y)$ are the magnitude and phase of its fundamental mode and $\Delta R_j(x, y)$ and $\Phi_j(x, y)$ $(j = 2, 3, \ldots)$ are the magnitude and phase of its jth harmonic.

In this chapter, we explain and express mathematically the relationship between the perturbed refractive index of Eq. (2.1) and the perturbed reflectance of Eq. (2.2). In order to do so, two physical phenomena, i.e. *reflection* and *optical interference*, as well as their causes need to be introduced.

First, *reflection* occurs when an electromagnetic wave reaches an interface between two media with different complex refractive indices (see Fig. 2.1). The boundary conditions of Maxwell's wave equations indeed show that only a proportion $\tilde{t}$ of the electromagnetic wave is transmitted through the interface. The rest of it, i.e. a proportion $\tilde{r} = \tilde{t} - 1$, bounces back or *is reflected* in opposite direction [1].

In other words, for an incident electric field $\tilde{E}_i$, the transmitted electric field is $\tilde{E}_t = \tilde{t}\tilde{E}_i$ and the reflected electric field is $\tilde{E}_r = \tilde{r}\tilde{E}_i$, where $\tilde{t}$ and $\tilde{r}$ are respectively called the transmission and reflection coefficients. Mathematically, the reflection coefficient $\tilde{r}$ of a light beam normally incident upon an interface separating two media with different complex refractive indices $\tilde{n}_1$ and $\tilde{n}_2$ follows Fresnel's reflection formula [1], i.e.

$$\tilde{r} = \frac{\tilde{n}_1 - \tilde{n}_2}{\tilde{n}_1 + \tilde{n}_2}. \tag{2.3}$$

The electromagnetic wave is therefore transmitted with a transmission coefficient $\tilde{t}$ such that

$$\tilde{t} = 1 + \tilde{r} = \frac{2\tilde{n}_1}{\tilde{n}_1 + \tilde{n}_2}. \tag{2.4}$$

The reflectance of the sample is then given by the squared absolute value of the reflection coefficient, i.e.

$$R = |\tilde{r}|^2 . \tag{2.5}$$

Second, *optical interference* is the name given to the interaction between two (or more) coherent light beams of the same optical frequency and polarization meeting in a region of space [2]. In summary, the interference describes the peculiar way these waves add up so as to give a total amplitude with is *not* simply the sum of their amplitudes.

Mathematically, the electric field of a one-dimensional electromagnetic plane wave in a medium of refractive index $\tilde{n}$ can be described as follows [1]

$$\tilde{E} = |E| \exp(2i\pi\tilde{n}x/\lambda_E + i\theta_E) \exp(-i\omega_E t), \tag{2.6}$$

where $|E|$ is the amplitude of the electric field (its maximum value in time and space), θ_E is its phase, λ_E is its wavelength in vacuum and ω_E is its optical angular frequency. If two electromagnetic waves of the same amplitude and different phases $\tilde{E}_1 = |E| \exp(2i\pi\tilde{n}x/\lambda_E + i\theta_{E1}) \exp(-i\omega_E t)$ and $\tilde{E}_2 = |E| \exp(2i\pi\tilde{n}x/\lambda_E + i\theta_{E2}) \exp(-i\omega_E t)$ meet, their amplitudes add up *vectorially* (or *coherently*) such that

$$\tilde{E}_1 + \tilde{E}_2 = |E| \left[\exp(i\theta_{E1}) + \exp(i\theta_{E2})\right] \exp(2i\pi\tilde{n}x/\lambda_E - i\omega_E t). \tag{2.7}$$

Equation (2.7) shows that the resulting amplitude is not solely dependent on the amplitude of the two beams but that it also strongly depends on their phase difference. If the two waves have the same phase (i.e. $\theta_{E1} = \theta_{E2}$), the amplitude of their sum is $2\,|E|$, i.e. there is *constructive interference*. On the contrary if the phases of the two waves are in opposition (i.e. $\theta_{E1} = \theta_{E2} + \pi$), the sum is zero, giving rise to *destructive interference*. For other values of the phase difference, the sum will be included between 0 and $2\,|E|$. Similarly, in the more general case of waves with different amplitudes $|E_1|$ and $|E_2|$, destructive and constructive interferences result in total fields of respective amplitudes $|E_1 - E_2|$ and $|E_1 + E_2|$.

One common case of interference arises when a light beam is reflected on two or more parallel interfaces. The case of two interfaces is well known and widely studied under the name of thin-film interference [2] (Fig. 2.2). In that case, $\tilde{E}_1$ is the electric field reflected directly on the top surface (Interface 1) and $\tilde{E}_2$ is the electric field reflected on the bottom surface (Interface 2). Due to distance traveled by $\tilde{E}_2$ in between the two interfaces, the phase difference between the two reflected electric fields is proportional to the thickness of the film. The combination of reflection and interference therefore offers a very attractive sensitivity to the thickness of a thin film. This is what makes reflection techniques so appealing. This experiment is studied in further detail in Sect. 2.2.

Due to interference effects, the mathematical relationships between the components of $\tilde{n}(x, y, z, t)$ and $R(x, y, t)$ vary according to the depth- (i.e. the z-) dependence of the refractive index perturbation. In this chapter, we investigate four different cases of refractive index perturbations and work out these relationships in order of

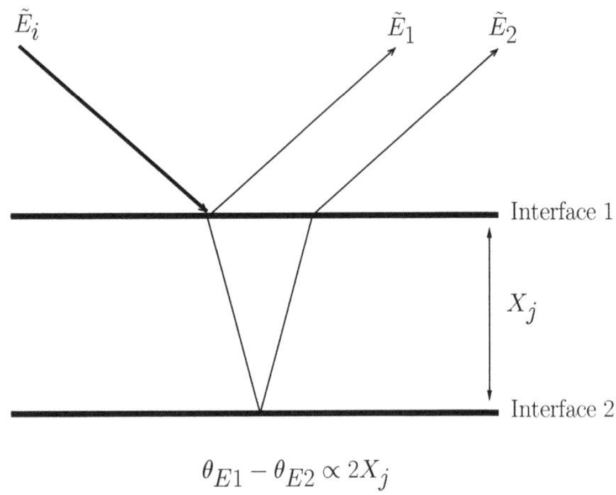

Fig. 2.2 Thin-film interference: the electric fields $\tilde{E}_1$ and $\tilde{E}_2$ respectively reflected on interface 1 and interface 2 present a phase difference $\theta_{E1} - \theta_{E2}$ due to the distance traveled by the light in between the two interfaces. As a consequence, the amplitude of the total reflected electric field varies with the thickness of the thin film

increasing complexity. First, uniform perturbations are studied in Sect. 2.1. In this case, the complex refractive index varies only at the top surface. Second, in Sect. 2.2, we consider the case of a box-like perturbation where the perturbed refractive index shows two abrupt variations, one at the top surface and one at a depth X_j, which we call the junction depth. Finally, we study the case of a double box-like perturbation (three abrupt variations) in Sect. 2.3 before deriving a formula for a perturbation with an arbitrary depth dependence in Sect. 2.4.

These analytical expressions are here derived in the case of TP, i.e. for a probe laser in the red to near infrared (NIR) range normally incident on the sample and for a system lying in air. As a consequence, the penetration depth of the probe laser $1/\alpha_{\text{probe}}$ is always much larger than the total depth of the perturbation of the complex refractive index (ultra-shallow perturbations) but much shorter than the thickness of the sample (semi-infinite samples). Further, we make the following three key assumptions so as to keep the expressions simple and analytical. First, the Si surface is supposed to be oxide-free, i.e. silicon is in direct contact with air. Second, we consider only laterally homogeneous refractive index perturbations, i.e. $\Delta\tilde{n}_j(x, y, z) = \Delta\tilde{n}_j(z)$ ($j = 0, 1, \ldots$). These first two assumptions are discussed into more detail at the end of this chapter respectively in Sects. 2.5.1 and 2.5.2. Finally, all the perturbation components of Eq. (2.1) are assumed much smaller than $\tilde{n}_0$. This linearizes the system (no harmonic mixing) and therefore greatly simplifies the obtained expressions. In other words, only a DC perturbation $\Delta\tilde{n}_0$ of the complex refractive index can induce a DC variation ΔR_0 in reflectance. Similarly, only the fundamental mode of the complex refractive index can generate ΔR_1. Besides, the relationships between the perturbation components of $\tilde{n}$ and R are the same independently from the considered component. We can therefore focus on a case of a single perturbation, relying on the linearity of the system for the cases of multiple perturbations. The j subscript ($j = 0, 1, \ldots$) is therefore omitted in the rest of this chapter.

For the sake of completeness, let us mention that two formulations are available to solve the considered problem. We propose to solve this problem in the *small*

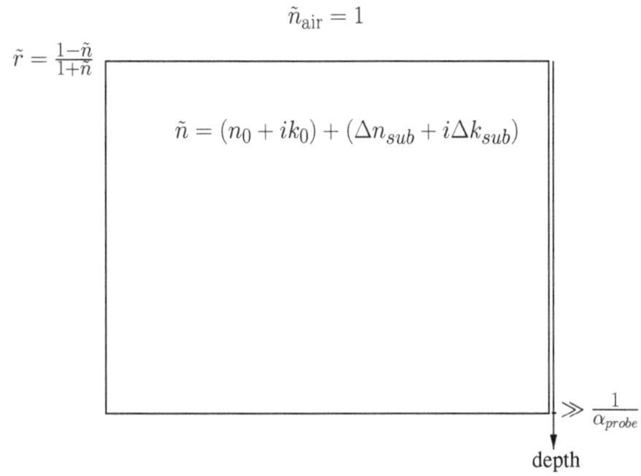

Fig. 2.3 A uniform perturbation $(\Delta n_{sub} + i\Delta k_{sub})$ of the refractive index $(n_0 + ik_0)$ of a semi-infinite sample

perturbation formalism in this chapter. The alternative formulation, i.e. the direct *differentiation*, was derived by Seraphin [3] and Aspnes [4, 5]. With some algebra, it can be shown that the two formulations are equivalent.

2.1 Uniform Perturbation of the Complex Refractive Index

In this Section, we consider a sample with a complex refractive index $(n_0 + ik_0)$ modified by a uniform perturbation $(\Delta n_{sub} + i\Delta k_{sub})$ and would like to calculate how this perturbation impacts the sample reflectance. The studied situation is depicted in Fig. 2.3.

In the case of a perturbed complex refractive index, Fresnel's formula (2.3) becomes

$$\tilde{r} = \frac{1 - (n_0 + ik_0) - (\Delta n_{sub} + i\Delta k_{sub})}{1 + (n_0 + ik_0) + (\Delta n_{sub} + i\Delta k_{sub})}. \tag{2.8}$$

Linearizing formula (2.8) with respect to the perturbation, $\tilde{r}$ becomes

$$\begin{aligned}
\tilde{r} &\approx \underbrace{\frac{1 - (n_0 + ik_0)}{1 + (n_0 + ik_0)}}_{=\tilde{r}_0} \left[\left(1 - \frac{\Delta n_{sub} + i\Delta k_{sub}}{1 - n_0 - ik_0}\right)\left(1 - \frac{\Delta n_{sub} + i\Delta k_{sub}}{1 + n_0 + ik_0}\right)\right] \\
&\approx \tilde{r}_0 \left[1 - \frac{2}{(1 - n_0 - ik_0)(1 + n_0 + ik_0)}(\Delta n_{sub} + i\Delta k_{sub})\right] \\
&= \tilde{r}_0 \left\{1 - \frac{2\left[(1 - n_0^2 + k_0^2)\Delta n_{sub} - 2n_0k_0\Delta k_{sub}\right]}{(1 - n_0^2 + k_0^2)^2 + 4n_0^2k_0^2} \right. \\
&\qquad \left. - i\frac{2\left[(1 - n_0^2 + k_0^2)\Delta k_{sub} + 2n_0k_0\Delta n_{sub}\right]}{(1 - n_0^2 + k_0^2)^2 + 4n_0^2k_0^2}\right\},
\end{aligned} \tag{2.9}$$

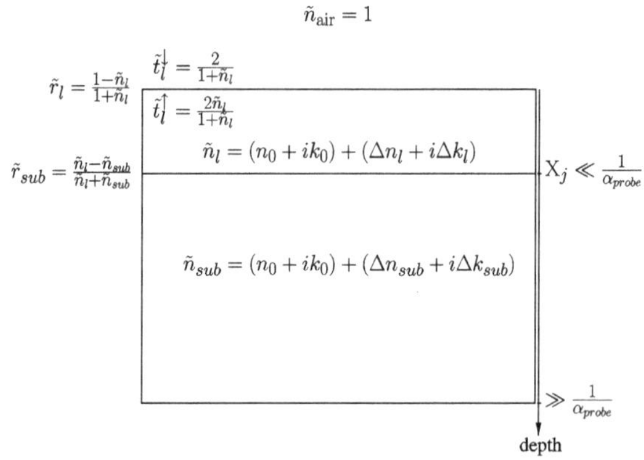

Fig. 2.4 A box-like perturbation of the refractive index profile shows two abrupt variations, respectively at the surface and at the interface. The interface is located at a depth X_j assumed to be much smaller than the penetration depth of the probe laser $(1/\alpha_{\text{probe}})$

where $\tilde{r}_0$ is the reflection coefficient of the unperturbed sample. The perturbed reflectance $R = |\tilde{r}|^2$ is therefore

$$R = \underbrace{|\tilde{r}_0|^2}_{=R_0} \left\{ 1 - \frac{4}{(1-n_0^2+k_0^2)^2 + 4n_0^2k_0^2} \left[(1-n_0^2+k_0^2)\Delta n_{sub} - 2n_0k_0\Delta k_{sub} \right] \right\}. \tag{2.10}$$

Before concluding this Section, it is interesting to note that $\partial R/\partial \Delta k_{sub}$ is proportional to k_0, which is very small in silicon in the red and NIR range [6]. In other words, a uniform Δk_{sub} hardly perturbs the reflectance of silicon in the red and NIR range. Further neglecting all the k_0 terms of Eq. (2.10), the variation ΔR in reflectance for a homogeneous perturbation simply reads

$$\Delta R|_{\text{homogeneous}} = \frac{4R_0}{n_0^2 - 1}\Delta n_{sub}. \tag{2.11}$$

This formula will prove very helpful for calculating the perturbation of the reflectance on a homogeneously doped sample, whether it is due to homogeneous doping, optically injected carriers or heat (Sect. 6.1).

2.2 Box-Like Perturbation of the Complex Refractive Index

We consider here the problem of a perturbation of the refractive index which only shows two abrupt transitions, one at the top surface and one at a depth X_j, called the junction depth. The perturbation of the refractive index has a value $(\Delta n_l + i\Delta k_l)$ in the box and a value $(\Delta n_{sub} + i\Delta k_{sub})$ below the layer. The unperturbed refractive index of the sample is uniform and equal to $(n_0 + ik_0)$. This situation is described in Fig. 2.4. We recognize here the thin-film interference introduced earlier in the introduction of the present chapter.

If we assume that the magnitude of the perturbation of refractive index is too small to cause multireflections in the box [2], the pertubation of the reflection coefficient is solely due to the *coherent* sum of the two reflections occurring respectively at the surface and at the interface. Further, neglecting the impact of the refractive index perturbation on the phase of the transmitted electric field, the reflection coefficient is

$$\tilde{r} = \tilde{r}_l + \tilde{r}_{sub}\tilde{t}^{\uparrow}\tilde{t}^{\downarrow}\exp(4i\pi n_0 X_j/\lambda_{\text{probe}}), \tag{2.12}$$

where $\tilde{r}_l$ and $\tilde{r}_{sub}$ are the reflection coefficients respectively at the surface and the interface, $\tilde{t}_l^{\downarrow}$ and $\tilde{t}_l^{\uparrow}$ are the transmission coefficients through the surface respectively for incoming and outgoing light.

In analogy to formula (2.9), we have for $\tilde{r}_l$

$$\begin{aligned}\tilde{r}_l = \tilde{r}_0\Bigg\{1 &- \frac{2\left[(1-n_0^2+k_0^2)\Delta n_l - 2n_0k_0\Delta k_l\right]}{(1-n_0^2+k_0^2)^2+4n_0^2k_0^2}\\ &- i\frac{2\left[(1-n_0^2+k_0^2)\Delta k_l + 2n_0k_0\Delta n_l\right]}{(1-n_0^2+k_0^2)^2+4n_0^2k_0^2}\Bigg\}.\end{aligned} \tag{2.13}$$

As for $\tilde{r}_{sub}$, neglecting all second-order perturbation terms, one obtains

$$\begin{aligned}\tilde{r}_{sub} &= \frac{(\Delta n_l + i\Delta k_l) - (\Delta n_{sub} + i\Delta k_{sub})}{2(n_0+ik_0) + (\Delta n_l + i\Delta k_l) + (\Delta n_{sub} + i\Delta k_{sub})}\\ &\approx \frac{(\Delta n_l - \Delta n_{sub}) + i(\Delta k_l - \Delta k_{sub})}{2(n_0+ik_0)}.\end{aligned} \tag{2.14}$$

It is apparent from the comparison of Eqs. (2.13) and (2.14) that the surface and interface reflections have very different magnitudes. While the surface reflection is composed of two terms, respectively of the zeroth and first orders, the interface reflection only shows a first-order contribution. This is due to the fact that, unlike the surface reflection, the interface reflection only exists because of the perturbation. Since we neglect all second-order terms, using formula (2.4), we can write

$$\begin{aligned}\tilde{r}_{sub}\tilde{t}^{\uparrow}\tilde{t}^{\downarrow} &= 2\tilde{r}_0\frac{(\Delta n_l - \Delta n_{sub}) + i(\Delta k_l - \Delta k_{sub})}{(1-n_0-ik_0)(1+n_0+ik_0)}\\ &= 2\tilde{r}_0\Bigg\{\frac{(1-n_0^2+k_0^2)(\Delta n_l - \Delta n_{sub}) - 2k_0n_0(\Delta k_l - \Delta k_{sub})}{(1-n_0^2+k_0^2)^2+4n_0^2k_0^2}\\ &\quad + i\frac{(1-n_0^2+k_0^2)(\Delta k_l - \Delta k_{sub}) + 2k_0n_0(\Delta n_l - \Delta n_{sub})}{(1-n_0^2+k_0^2)^2+4n_0^2k_0^2}\Bigg\}.\end{aligned} \tag{2.15}$$

Putting Eqs. (2.12), (2.13) and (2.15) together and neglecting all second-order terms, this gives for the reflectance

$$
\begin{aligned}
R &\approx |\tilde{r}_l|^2 + 2\Re(\tilde{r}_0^\star \tilde{r}_{sub}\tilde{t}^\uparrow \tilde{t}^\downarrow \exp(4i\pi n_0 X_j/\lambda_{\text{probe}})) \\
&= R_0\Bigg\{1 - \frac{4}{(1-n_0^2+k_0^2)^2+4n_0^2k_0^2} \\
&\times \Bigg[(1-n_0^2+k_0^2)\Delta n_l - 2n_0k_0\Delta k_l \\
&- \Big((1-n_0^2+k_0^2)(\Delta n_l - \Delta n_{sub}) - 2k_0n_0(\Delta k_l - \Delta k_{sub})\Big) \\
&\times \cos(4\pi n_0 X_j/\lambda_{\text{probe}}) \\
&+ \Big((1-n_0^2+k_0^2)(\Delta k_l - \Delta k_{sub}) + 2k_0n_0(\Delta n_l - \Delta n_{sub})\Big) \\
&\times \sin(4\pi n_0 X_j/\lambda_{\text{probe}})\Bigg]\Bigg\},
\end{aligned} \tag{2.16}
$$

where $\tilde{r}_0^\star$ is the complex conjugate of $\tilde{r}_0$. Assuming again that $k_0 \ll n_0$, the perturbation of the reflectance is

$$
\begin{aligned}
\Delta R|_{\text{box}} = \frac{4R_0}{n_0^2-1}\big[&\Delta n_l \\
&- \cos(4\pi n_0 X_j/\lambda_{\text{probe}})(\Delta n_l - \Delta n_{sub}) \\
&+ \sin(4\pi n_0 X_j/\lambda_{\text{probe}})(\Delta k_l - \Delta k_{sub})\big].
\end{aligned} \tag{2.17}
$$

This equation is of very high importance in this work. It indeed shows the interest of reflection techniques for the depth-determination of a refractive index perturbation. Though reflection techniques are surface techniques, they are able to probe the in-depth variations in refractive index with high sensitivity thanks to the interference between the reflections respectively occuring at the surface and at the interface of the box. This is the main reason why optical reflection techniques are usually very attractive for the non-destructive determination of e.g. layer thicknesses [2]. Besides, Eq. (2.17) proves to explain with great accuracy the perturbed reflectances experimentally observed on CVD box-like doping profiles (Chap. 6.2). It will therefore be frequently used for the assessment of our model.

Notice finally that, if X_j=0, formula (2.17) nicely reduces to formula (2.11).

2.3 Double Box-Like Perturbation of the Complex Refractive Index

We consider here the problem of a perturbation of the refractive index which has three abrupt transitions, one at the top surface, one at a depth X_{j1} and a last one at a depth X_{j2}. This situation is shown in Fig. 2.5. The perturbation of the refractive index has a value $(\Delta n_{l1} + i\Delta k_{l1})$ in the top box, a value $(\Delta n_{l2} + i\Delta k_{l2})$ in the second

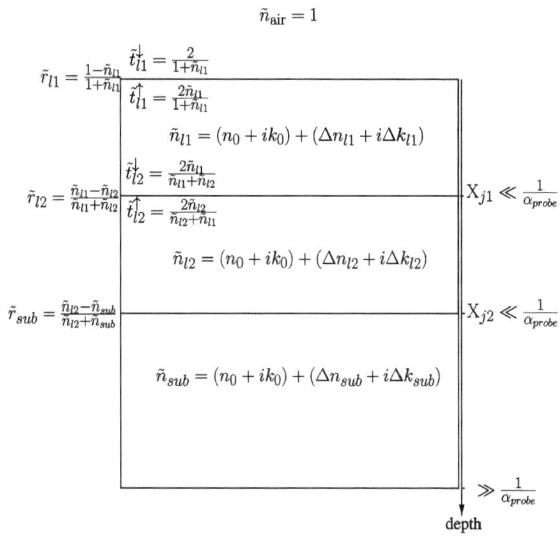

Fig. 2.5 A double box-like perturbation of the refractive index profile shows three abrupt variations, respectively at the surface, at a depth X_{j1} and at a depth X_{j2}

box and a value $(\Delta n_{sub} + i\,\Delta k_{sub})$ below the second box. The unperturbed refractive index of the sample is uniform and equal to $(n_0 + ik_0)$.

As can be expected, the perturbation of the reflection coefficient is here due to the coherent sum of the *three* reflections occurring at the surface and at the two interfaces. In other words, generalizing Eq. (2.12), the reflection coefficient can be written

$$\tilde{r} = \tilde{r}_{l1} + \tilde{r}_{l2}\tilde{t}^{\uparrow}_{l1}\tilde{t}^{\downarrow}_{l1}\exp(4i\pi n_0 X_{j1}/\lambda_{\rm probe}) + \tilde{r}_{sub}\tilde{t}^{\uparrow}_{l1}\tilde{t}^{\downarrow}_{l1}\tilde{t}^{\uparrow}_{l2}\tilde{t}^{\downarrow}_{l2}\exp(4i\pi n_0 X_{j2}/\lambda_{\rm probe}). \tag{2.18}$$

In analogy with Eq. (2.9) for $\tilde{r}_{l1}$ and Eq. (2.14) for $\tilde{r}_{l2}$ and $\tilde{r}_{sub}$, the reflection coefficients on each interface are respectively

$$\tilde{r}_{l1} = \tilde{r}_0\left\{1 - \frac{2\left[(1-n_0^2+k_0^2)\Delta n_{l1} - 2n_0k_0\Delta k_{l1}\right]}{(1-n_0^2+k_0^2)^2 + 4n_0^2k_0^2} - i\frac{2\left[(1-n_0^2+k_0^2)\Delta k_{l1} + 2n_0k_0\Delta n_{l1}\right]}{(1-n_0^2+k_0^2)^2 + 4n_0^2k_0^2}\right\} \tag{2.19}$$

$$\tilde{r}_{l2} = \frac{(\Delta n_{l1} - \Delta n_{l2}) + i(\Delta k_{l1} - \Delta k_{l2})}{2(n_0 + ik_0)} \tag{2.20}$$

$$\tilde{r}_{sub} = \frac{(\Delta n_{l2} - \Delta n_{sub}) + i(\Delta k_{l2} - \Delta k_{sub})}{2(n_0 + ik_0)}. \tag{2.21}$$

Similarly to Eq. (2.15), we have

$$\tilde{r}_{l2}\tilde{t}_{l1}^{\uparrow}\tilde{t}_{l1}^{\downarrow} = 2\tilde{r}_0 \left\{ \frac{(1-n_0^2+k_0^2)(\Delta n_{l1}-\Delta n_{l2}) - 2k_0 n_0(\Delta k_{l1}-\Delta k_{l2})}{(1-n_0^2+k_0^2)^2+4n_0^2k_0^2} + i\frac{(1-n_0^2+k_0^2)(\Delta k_{l1}-\Delta k_{l2}) + 2k_0 n_0(\Delta n_{l1}-\Delta n_{l2})}{(1-n_0^2+k_0^2)^2+4n_0^2k_0^2} \right\} \quad (2.22)$$

$$\tilde{r}_{sub}\tilde{t}_{l1}^{\uparrow}\tilde{t}_{l1}^{\downarrow}\tilde{t}_{l2}^{\uparrow}\tilde{t}_{l2}^{\downarrow} = 2\tilde{r}_0 \left\{ \frac{(1-n_0^2+k_0^2)(\Delta n_{l2}-\Delta n_{sub}) - 2k_0 n_0(\Delta k_{l2}-\Delta k_{sub})}{(1-n_0^2+k_0^2)^2+4n_0^2k_0^2} + i\frac{(1-n_0^2+k_0^2)(\Delta k_{l2}-\Delta k_{sub}) + 2k_0 n_0(\Delta n_{l2}-\Delta n_{sub})}{(1-n_0^2+k_0^2)^2+4n_0^2k_0^2} \right\}. \quad (2.23)$$

Plugging Eqs. (2.19), (2.22) and (2.23) into Eq. (2.18) and neglecting all second-order terms in refractive index variations, we have

$$\begin{aligned} R &= |\tilde{r}_{l1}|^2 \\ &\quad + 2\Re(\tilde{r}_0^\star \tilde{r}_{l2}\tilde{t}_{l1}^{\uparrow}\tilde{t}_{l1}^{\downarrow}\exp(4i\pi n_0 X_{j1}/\lambda_{\text{probe}})) \\ &\quad + 2\Re(\tilde{r}_0^\star \tilde{r}_{sub}\tilde{t}_{l1}^{\uparrow}\tilde{t}_{l1}^{\downarrow}\tilde{t}_{l2}^{\uparrow}\tilde{t}_{l2}^{\downarrow}\exp(4i\pi n_0 X_{j2}/\lambda_{\text{probe}})) \\ &= R_0 \Bigg\{ 1 - \frac{4}{(1-n_0^2+k_0^2)^2+4n_0^2k_0^2} \\ &\quad \times \Bigg[(1-n_0^2+k_0^2)\Delta n_{l1} - 2n_0 k_0 \Delta k_{l1} \\ &\quad - \Big((1-n_0^2+k_0^2)(\Delta n_{l1}-\Delta n_{l2}) - 2k_0 n_0(\Delta k_{l1}-\Delta k_{l2})\Big) \\ &\quad \times \cos(4\pi n_0 X_{j1}/\lambda_{\text{probe}}) \\ &\quad + \Big((1-n_0^2+k_0^2)(\Delta k_{l1}-\Delta k_{l2}) + 2k_0 n_0(\Delta n_{l1}-\Delta n_{l2})\Big) \\ &\quad \times \sin(4\pi n_0 X_{j1}/\lambda_{\text{probe}}) \\ &\quad - \Big((1-n_0^2+k_0^2)(\Delta n_{l2}-\Delta n_{sub}) - 2k_0 n_0(\Delta k_{l2}-\Delta k_{sub})\Big) \\ &\quad \times \cos(4\pi n_0 X_{j2}/\lambda_{\text{probe}}) \\ &\quad + \Big((1-n_0^2+k_0^2)(\Delta k_{l2}-\Delta k_{sub}) + 2k_0 n_0(\Delta n_{l2}-\Delta n_{sub})\Big) \\ &\quad \times \sin(4\pi n_0 X_{j2}/\lambda_{\text{probe}}) \Bigg] \Bigg\}. \end{aligned} \quad (2.24)$$

Assuming again that $k_0 \ll n_0$, the perturbation of the reflectance is

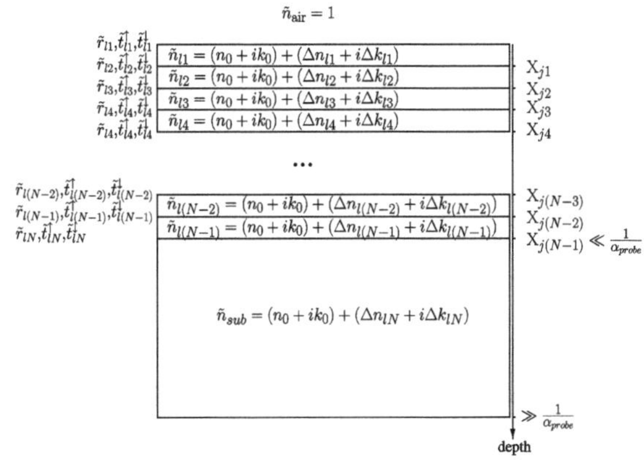

Fig. 2.6 A staircase perturbation of the refractive index profile with N abrupt transitions

$$
\begin{aligned}
\Delta R|_{\text{double-box}} = \frac{4R_0}{(n_0^2 - 1)} \Big[& \Delta n_{l1} \\
& - \cos(4\pi n_0 X_{j1}/\lambda_{\text{probe}})(\Delta n_{l1} - \Delta n_{l2}) \\
& - \cos(4\pi n_0 X_{j2}/\lambda_{\text{probe}})(\Delta n_{l2} - \Delta n_{sub}) \\
& + \sin(4\pi n_0 X_{j1}/\lambda_{\text{probe}})(\Delta k_{l1} - \Delta k_{l2}) \\
& + \sin(4\pi n_0 X_{j2}/\lambda_{\text{probe}})(\Delta k_{l2} - \Delta k_{sub}) \Big].
\end{aligned} \tag{2.25}
$$

2.4 Arbitrary Perturbation of the Complex Refractive Index

Building the theory as we have, it is now fairly easy to derive a general formula for a staircase perturbation of the complex refractive index with N abrupt transitions such as presented in Fig. 2.6.

Generalizing Eqs. (2.12) and (2.18) for the case of an N-transition staircase gives

$$
\begin{aligned}
\tilde{r} = {} & \tilde{r}_{l1} + \tilde{r}_{l2}\tilde{t}_{l1}^{\uparrow}\tilde{t}_{l1}^{\downarrow} \exp(4i\pi n_0 X_{j1}/\lambda_{\text{probe}}) + \tilde{r}_{l3}\tilde{t}_{l1}^{\uparrow}\tilde{t}_{l1}^{\downarrow}\tilde{t}_{l2}^{\uparrow}\tilde{t}_{l2}^{\downarrow} \exp(4i\pi n_0 X_{j2}/\lambda_{\text{probe}}) \\
& + \cdots \\
& + \tilde{r}_{l(N-1)}\tilde{t}_{l1}^{\uparrow}\tilde{t}_{l1}^{\downarrow}\tilde{t}_{l2}^{\uparrow}\tilde{t}_{l2}^{\downarrow} \ldots \tilde{t}_{l(N-2)}^{\uparrow}\tilde{t}_{l(N-2)}^{\downarrow} \exp(4i\pi n_0 X_{j(N-2)}/\lambda_{\text{probe}}) \\
& + \tilde{r}_{lN}\tilde{t}_{l1}^{\uparrow}\tilde{t}_{l1}^{\downarrow}\tilde{t}_{l2}^{\uparrow}\tilde{t}_{l2}^{\downarrow} \ldots \tilde{t}_{l(N-1)}^{\uparrow}\tilde{t}_{l(N-1)}^{\downarrow} \exp(4i\pi n_0 X_{j(N-1)}/\lambda_{\text{probe}}).
\end{aligned} \tag{2.26}
$$

The expression for the surface reflection $\tilde{r}_{l1}$ is given by Eq. (2.19). As for the other reflections $\tilde{r}_{l\eta}$ ($\eta = 2, 3, \ldots, N$), a generalization of Eq. (2.20) gives

$$
\tilde{r}_{l\eta} = \frac{(\Delta n_{l(\eta-1)} - \Delta n_{l\eta}) + i(\Delta k_{l(\eta-1)} - \Delta k_{l\eta})}{2(n_0 + ik_0)}. \tag{2.27}
$$

Generalizing Eq. (2.22), we therefore have

$$\tilde{r}_{l\eta}\tilde{t}^{\uparrow}_{l1}\tilde{t}^{\downarrow}_{l1}\dots\tilde{t}^{\uparrow}_{l(\eta-1)}\tilde{t}^{\downarrow}_{l(\eta-1)}$$
$$= 2\tilde{r}_0\left\{\frac{(1-n_0^2+k_0^2)(\Delta n_{l(\eta-1)}-\Delta n_{l\eta})-2k_0n_0(\Delta k_{l(\eta-1)}-\Delta k_{l\eta})}{(1-n_0^2+k_0^2)^2+4n_0^2k_0^2}\right.$$
$$\left.+i\frac{(1-n_0^2+k_0^2)(\Delta k_{l(\eta-1)}-\Delta k_{l\eta})+2k_0n_0(\Delta n_{l(\eta-1)}-\Delta n_{l\eta})}{(1-n_0^2+k_0^2)^2+4n_0^2k_0^2}\right\}. \quad (2.28)$$

which gives the reflectance when plugged into the following generalization of Eq. (2.24)

$$R = |\tilde{r}_{l1}|^2$$
$$+ 2\Re\left(\tilde{r}_0^{\star}\sum_{\eta=2}^{N}\tilde{r}_{l\eta}t^{\uparrow}_{l1}\tilde{t}^{\downarrow}_{l1}\dots t^{\uparrow}_{l(\eta-1)}\tilde{t}^{\downarrow}_{l(\eta-1)}\exp(4i\pi n_0X_{j(\eta-1)}/\lambda_{\text{probe}})\right). \quad (2.29)$$

If the thickness of the layers becomes infinitesimal, i.e. the perturbation of the refractive index varies continuously with depth, Eq. (2.27) becomes

$$\tilde{r}_{l\eta} = \frac{d(\Delta n+i\Delta k)}{2(n_0+ik_0)}$$
$$= \frac{1}{2(n_0+ik_0)}\frac{\partial(\Delta n+i\Delta k)}{\partial z}dz, \quad (2.30)$$

and the sum appearing in Eq. (2.29) must be turned into an integral to give

$$R = |\tilde{r}_{l1}|^2 + 2\Re\left(\tilde{r}_0^{\star}\frac{2\tilde{r}_0}{(1-n_0-ik_0)(1+n_0+ik_0)}\right.$$
$$\left.\times\int_{0+}^{+\infty}\frac{\partial(\Delta n+i\Delta k)}{\partial z}\exp(4i\pi n_0z/\lambda_{\text{probe}})dz\right)$$
$$= R_0\left\{1-\frac{4}{(1-n_0^2+k_0^2)^2+4n_0^2k_0^2}\times\left[(1-n_0^2+k_0^2)\Delta n(z=0)-2n_0k_0\Delta k(z=0)\right.\right.$$
$$+(1+n_0^2+k_0^2)\left(\int_{0+}^{+\infty}\frac{\partial\Delta n(z)}{\partial z}\cos(4\pi n_0z/\lambda_{\text{probe}})dz-\frac{\partial\Delta k(z)}{\partial z}\sin(4\pi n_0z/\lambda_{\text{probe}})dz\right)$$
$$\left.\left.+2k_0n_0\left(\int_{0+}^{+\infty}\frac{\partial\Delta k(z)}{\partial z}\cos(4\pi n_0z/\lambda_{\text{probe}})dz+\frac{\partial\Delta n(z)}{\partial z}\sin(4\pi n_0z/\lambda_{\text{probe}})dz\right)\right]\right\}. \quad (2.31)$$

For Si in the red and NIR range, Eq. (2.31) becomes

$$\Delta R|_{\text{Profile}} = \frac{4R_0}{(n_0^2-1)}\left[\Delta n(z=0) + \int_{0+}^{+\infty}\left(\frac{\partial \Delta n(z)}{\partial z}\cos(4\pi n_0 z/\lambda_{\text{probe}}) - \frac{\partial \Delta k(z)}{\partial z}\sin(4\pi n_0 z/\lambda_{\text{probe}})\right)dz\right]. \tag{2.32}$$

It can be easily checked that in the case of a box-like perturbation of the refractive index, formula (2.32) reduces to (2.17).

2.5 Second-Order Effects

As mentioned in the Introduction, the reflection formulas derived in this chapter assume the absence of any oxide layer at the interface between air and the silicon sample. Similarly, the derived formulas only consider laterally homogeneous perturbations of the refractive index. These two assumptions are respectively looked at in Sects. 2.5.1 and 2.5.2 below.

2.5.1 Impact of the Presence of a Native Oxide

An oxide always exists at the interface between air and silicon. Hence, its impact should be evaluated. For this purpose, we propose to consider the case of a box-like perturbation of the refractive index, as studied in Sect. 2.2. On top of the structure of Fig. 2.4, we add an oxide layer of thickness t_{oxide} and of refractive index[1] $n_{\text{oxide}} = 1.45$ [7], such as represented in Fig. 2.7.

This effect should be studied as a function of t_{oxide} but, since the native oxides present on our samples have measured thicknesses between 0 and 2 nm, we take the worst-case scenario, i.e $t_{\text{oxide}} = 2$ nm. We further assume that the refractive index perturbation does not propagate into the oxide. Figure 2.8 shows that, for a real refractive index perturbation with $\Delta n_{sub} = -10^{-4}$ (typical value encountered in the present work) and $\lambda_{\text{probe}} = 670$ nm, the normalized variation $|(\Delta R_{\text{Box}}^{\text{Oxide}} - \Delta R_{\text{Box}})/\Delta R_{\text{Box}}(X_j = 0)|$ in reflectance perturbation is no more than 1.2 % whether $\Delta n_l = 0$ or -0.5×10^{-4}.

[1] Given the large refractive index contrast both at the air-oxide interface and at the oxide-silicon interface, multireflection formula [2] must be considered in the oxide layer (large reflection coefficients at both interfaces)

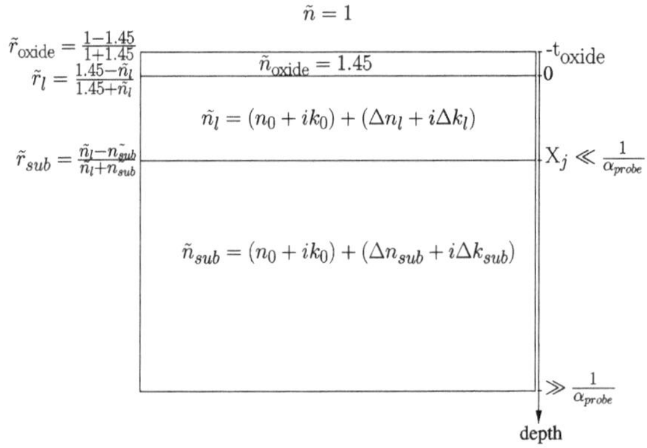

Fig. 2.7 Oxide layer of thickness t_{oxide} on top of a box-like perturbation of the refractive index profile

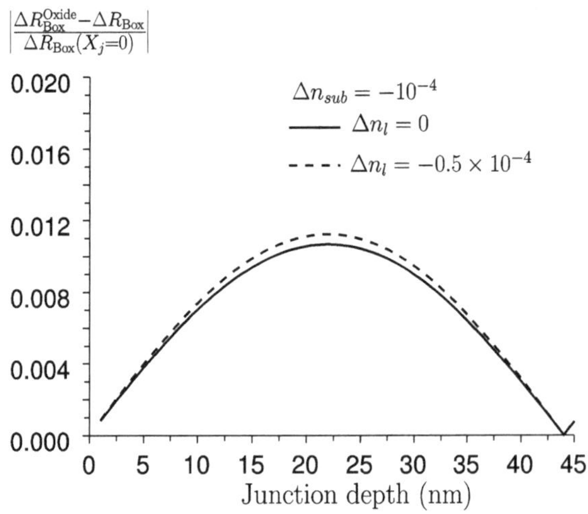

Fig. 2.8 Normalized variation $|(\Delta R_{\text{Box}}^{\text{Oxide}} - \Delta R_{\text{Box}})/\Delta R_{\text{Box}}(X_j = 0)|$ in reflectance perturbation due to the presence of a 2-nm thick oxide layer as a function of the junction depth of the box-like perturbation of the refractive index. The considered real values of the refractive index perturbation are indicated. The presence of the 2-nm thick native only has a negligible impact on the reflectance perturbation

For completeness, note that Fig. 2.8 does not show the behavior of the relative error $(\Delta R_{\text{Box}}^{\text{Oxide}} - \Delta R_{\text{Box}})/\Delta R_{\text{Box}}$. This error indeed diverges at $X_j \approx \lambda_{\text{probe}}/(8n_0) \approx 22$ nm, due to the vanishing amplitude of ΔR_{Box}. As Fig. 2.8 shows, the absolute error remains small even at $X_j \approx 22$ nm.

In summary, this effect can be neglected. It is, however, important to keep in mind that the variation grows with increasing oxide thickness.

2.5.2 Impact of a Lateral Variation in Refractive Index Perturbation

Thus far in this chapter, only laterally homogeneous perturbations of the refractive index have been considered. If $\Delta\tilde{n}(x, y, z)$ also varies with x and y, this effect will obviously be mirrored on the local reflectance perturbation $\Delta R(x, y)$. Typically, in this work, reflectance perturbations radially decay as damped waves (see Chap. 6), i.e.

$$\Delta R(r) = \Delta R(r = 0) \exp(-r/L_d^R) \cos(2\pi r/\Lambda^R), \tag{2.33}$$

where $r = \sqrt{x^2 + y^2}$ is defined as the radial distance to the maximum of the perturbation, L_d^R is the decay length and Λ^R is the wavelength of the reflectance perturbation. Figure 2.9a shows three examples of (normalized) reflectance perturbations behaving like damped waves with decay length $L_d^R = 2\,\mu\text{m}$ and respective wavelengths $\Lambda^R = 100, 10$ and $1\,\mu\text{m}$. As shown in Chap. 6, this is a relevant range for our experiments.

The probe laser irradiance distribution $\Pi_{\text{probe}}(r)$ being

$$\Pi_{\text{probe}}(r) = \mathcal{P}^0_{\text{probe}} \exp\left(-\frac{r^2}{\mathcal{R}^2_{\text{probe}}}\right), \tag{2.34}$$

where $\mathcal{P}^0_{\text{probe}}$ is the peak irradiance of the probe laser and $\mathcal{R}_{\text{probe}}$ is its radius, the measured reflectance perturbation $\Delta R_{\text{integrated}}$ will be a convolution of the local reflectance perturbation $\Delta R(r)$ with the probe laser irradiance distribution, i.e.

$$\begin{aligned}\Delta R_{\text{integrated}} &= \frac{2}{\mathcal{R}^2_{\text{laser}}} \int_0^\infty r dr \Delta R(r) \exp\left(-\frac{r^2}{\mathcal{R}^2_{\text{laser}}}\right) \\ &= \frac{2}{\mathcal{R}^2_{\text{laser}}} \int_0^\infty r dr \Delta R(r = 0) \exp(-r/L_d^R) \cos(2\pi r/\Lambda^R) \exp\left(-\frac{r^2}{\mathcal{R}^2_{\text{laser}}}\right),\end{aligned} \tag{2.35}$$

Note that we willingly restrict the present study to the case of corresponding positions of the maximum of the reflectance perturbation and of the probe irradiance.

Figure 2.9b shows the ratio of $\Delta R_{\text{integrated}}$, i.e. the reflectance perturbation as measured by the probe laser, divided by the local reflectance perturbation at $r = 0$, i.e. $\Delta R(r = 0)$. Obviously, when the decay length and wavelength are both long, $\Delta R_{\text{integrated}}$ is very close to $\Delta R(r = 0)$. This is the situation encountered in a vast majority ($> 95\,\%$) of our measurements, as highlighted by the dotted box in Fig. 2.9b. On the contrary, if the decay length and/or the wavelength become commensurate with $\mathcal{R}_{\text{probe}}$, strong deviations are observed, which can even lead to $\Delta R_{\text{integrated}}$

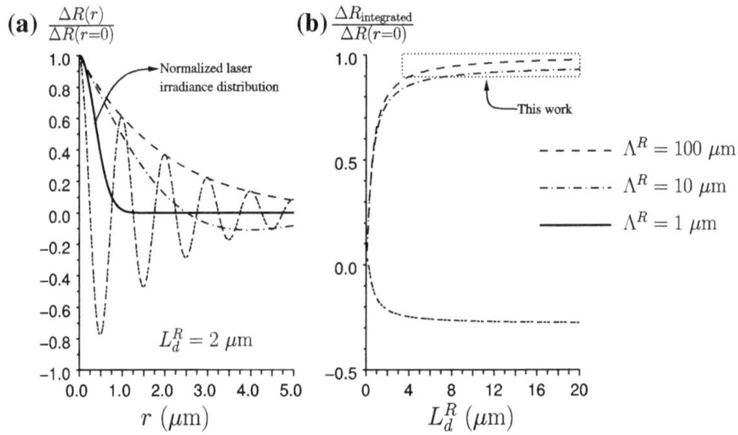

Fig. 2.9 **a** Examples of normalized reflectance perturbations behaving like damped waves (Eq. (2.33)) with decay length $L_d^R = 2\,\mu m$ and $\Lambda^R = 100\,\mu m$ (*dashed line*), $\Lambda^R = 10\,\mu m$ (*interrupted line*) and $\Lambda^R = 1\,\mu m$ (*triple-interrupted line*). The thick *black line* is the normalized irradiance distribution (Eq. (2.34) with $\mathcal{R}_{\text{probe}} = 0.5\,\mu m$). **b** Variation in the ratio of the integrated reflectance perturbation (Eq. (2.35)) divided by the signal at $r = 0$ as a function of the decay length of the reflectance perturbation. The dotted box shows the typical situations found in this work, where the impact of the lateral integration is minor

and $\Delta R(r = 0)$ being of opposite sign ($\Lambda^R = 1\,\mu m$). In these cases, the lateral integration cannot be ignored.

In this study, we try to work with analytical expressions as much as possible. The lateral integration of Eq. (2.35) will therefore be omitted in our model. If required, however, our calculations can take it into account. Only minor impact of this effect has been observed for the specific case of the structures measured in this work. This justifies the omission in the rest of this dissertation.

2.6 Summary

In this chapter, we have investigated the mathematical relationship between a refractive index perturbation and the subsequent perturbation of the sample reflectance, as can be measured by a probe laser in the red and NIR range. We have shown that, due to interference effects, the relationships vary greatly according to the depth-dependence of the refractive index perturbation. Of particular importance in this work, we have derived analytical expressions in the cases of a homogeneous refractive index perturbation

$$\Delta R|_{\text{homogeneous}} = \frac{4R_0}{n_0^2 - 1}\Delta n_{sub}, \tag{2.36}$$

in the case of a box-like refractive index perturbation

$$\Delta R|_{\text{box}} = \frac{4R_0}{n_0^2 - 1}\big[\Delta n_l - \cos(4\pi n_0 X_j/\lambda_{\text{probe}})(\Delta n_l - \Delta n_{sub}) + \sin(4\pi n_0 X_j/\lambda_{\text{probe}})(\Delta k_l - \Delta k_{sub})\big], \quad (2.37)$$

and in the case of an arbitrary profile of refractive index perturbation

$$\Delta R|_{\text{Profile}} = \frac{4R_0}{(n_0^2 - 1)}\left[\Delta n(z=0) + \int_{0+}^{+\infty}\left(\frac{\partial \Delta n(z)}{\partial z}\cos(4\pi n_0 z/\lambda_{\text{probe}}) - \frac{\partial \Delta k(z)}{\partial z}\sin(4\pi n_0 z/\lambda_{\text{probe}})\right)dz\right], \quad (2.38)$$

all the symbols being as defined in this chapter.

Further, we have shown that these expressions are valid even in the presence of a thin native oxide and in the case of laterally varying refractive index perturbations provided the lateral variations are not too abrupt.

References

1. M. Born, E. Wolf, *Principles of Optics* (Pergamon, Oxford, 1970)
2. H. Fujiwara, *Spectroscopic Ellipsometry: Principles and Applications* (Wiley, Chichester, 2007)
3. B.O. Seraphin, N. Bottka, Band-structure analysis from electro-reflectance studies. Phys. Rev. **145**(2), 628–636 (1966). doi:10.1103/PhysRev.145.628
4. D.E. Aspnes, in *Handbook on Semiconductors: Modulation Spectroscopy*, ed. by T.S. Moss (North Holland, New York, 1980)
5. D.E. Aspnes, A. Frova, Influence of spatially dependent perturbations on modulated reflectance and absorption of solids (reprinted in solid-state commun, vol 7, pg 155–159, 1969). Solid State Commun. **88**(11–12), 1061–1065 (1993)
6. T. Yasuda, D.E. Aspnes, Optical-standard surfaces of single-crystal silicon for calibrating ellipsometers and reflectometers. Appl. Opt. **33**(31), 7435–7438 (1994)
7. R. Hull (ed.), *Properties of Crystalline Silicon*, Emis Datareviews Series (INSPEC, London, 1999a)

Chapter 3
Theory of Perturbation of the Refractive Index

In this Chapter, we consider that some perturbation of the free electron distribution $N(x, y, z, t)$, of the free hole distribution $P(x, y, z, t)$ and of the temperature distribution $T(x, y, z, t)$ is present within a silicon sample due to doping and/or optical injection. We investigate how these distributions modify the complex refractive index $\tilde{n}(x, y, z, t)$ of the sample.

Without any loss of generality, the perturbed electron, hole and temperature distributions respectively read

$$\begin{aligned} N(x, y, z, t) &= N_{\text{doping}}(x, y, z) + \Delta N(x, y, z, t) \\ &= N_{\text{doping}}(x, y, z) + \Delta N_0(x, y, z) \\ &\quad + \underbrace{2|\Delta N_1(x, y, z)| \cos(\omega t - \theta_{N1}(x, y, z))}_{\text{fundamental}} \\ &\quad + \sum_{j=2}^{+\infty} \underbrace{2|\Delta N_j(x, y, z)| \cos(j\omega t - \theta_{Nj}(x, y, z))}_{j\text{ th harmonic}} \end{aligned} \tag{3.1}$$

$$\begin{aligned} P(x, y, z, t) &= P_{\text{doping}}(x, y, z) + \Delta P(x, y, z, t) \\ &= P_{\text{doping}}(x, y, z) + \Delta P_0(x, y, z) \\ &\quad + \underbrace{2|\Delta P_1(x, y, z)| \cos(\omega t - \theta_{P1}(x, y, z)))}_{\text{fundamental}} \\ &\quad + \sum_{j=2}^{+\infty} \underbrace{2|\Delta P_j(x, y, z)| \cos(j\omega t - \theta_{Pj}(x, y, z))}_{j\text{ th harmonic}} \end{aligned} \tag{3.2}$$

$$\begin{aligned} T(x, y, z, t) &= T_0 + \Delta T_0(x, y, z) + \underbrace{2|\Delta T_1(x, y, z)| \cos(\omega t - \theta_{T1}(x, y, z))}_{\text{fundamental}} \\ &\quad + \sum_{j=2}^{+\infty} \underbrace{2|\Delta T_j(x, y, z)| \cos(j\omega t - \theta_{Tj}(x, y, z))}_{j\text{ th harmonic}}, \end{aligned} \tag{3.3}$$

J. Bogdanowicz, *Photomodulated Optical Reflectance*, Springer Theses,
DOI: 10.1007/978-3-642-30108-7_3,

where N_{doping} (resp. P_{doping}) and ΔN (resp. ΔP) are respectively the electron (resp. hole) concentrations due to the homogeneous doping of the sample and due to optical injection. T_0 is the environment temperature during the experiment, i.e. room temperature (300 K). $\Delta N_0(x, y, z)$, $\Delta P_0(x, y, z)$ and $\Delta T_0(x, y, z)$ are the time-independent components of the perturbations respectively of the free electron distribution, of the free hole distribution and of the temperature distribution. $|\Delta N_1(x, y, z)|$ [resp. $\theta_{N1}(x, y, z)$], $|\Delta P_1(x, y, z)|$ [resp. $\theta_{P1}(x, y, z)$] and $|\Delta T_1(x, y, z)|$ [resp. $\theta_{T1}(x, y, z)$] are the half-amplitudes[1] (resp. phases) of the fundamental mode of the perturbations respectively of the free electron distribution, of the free hole distribution and of the temperature distribution. $\Delta N_j(x, y, z)$ [resp. $\theta_{Nj}(x, y, z)$], $\Delta P_j(x, y, z)$ [resp. $\theta_{Pj}(x, y, z)$] and $\Delta T_j(x, y, z)$ [resp. $\theta_{Tj}(x, y, z)$] are the half-amplitudes (resp. phases) of the jth harmonic of the perturbations respectively of the free electron distribution, of the free hole distribution and of the temperature distribution.

As a consequence of these perturbations, the complex refractive index $\tilde{n}(x, y, z, t)$ reads

$$\tilde{n}(x, y, z, t) = \tilde{n}_0 + \Delta\tilde{n}_0(x, y, z) + \underbrace{\Delta\tilde{n}_1(x, y, z)\cos(\omega t - \phi_1(x, y, z))}_{\text{fundamental}} + \sum_{j=2}^{+\infty} \underbrace{\Delta\tilde{n}_j(x, y, z)\cos(j\omega t - \phi_j(x, y, z))}_{j\text{ th harmonic}}, \tag{3.4}$$

as already introduced in Chap. 2.

In this Chapter, we identify the various physical phenomena accounting for variations in refractive index so as to link mathematically the different components of the perturbation of the refractive index appearing in Eq. (3.4) to the components of the perturbations of N, P and T appearing in Eqs. (3.1), (3.2) and (3.3). It is important to note that, similarly to Chap. 2, we here limit ourselves to small perturbations of the refractive index. As a consequence, no frequency mixing occurs and a general theory of perturbation can be developed independently from the time dependence of the perturbation. In the rest of this Chapter, we therefore drop the j subscript and look at the variation $\Delta\tilde{n}$ in refractive index due to arbitrary perturbations ΔN, ΔP and ΔT. In particular, we show, in Sect. 3.1, that free carriers, electric field and temperature can impact the refractive index $\tilde{n}$ of Si. In Sect. 3.2, we study the electrooptical effects, i.e. the effects of carriers and electric field on $\tilde{n}$. In Sect. 3.3, we investigate the thermooptical effects, i.e. the impact of temperature on $\tilde{n}$.

[1] $\Delta N_j(x, y, z)$, $\Delta P_j(x, y, z)$ and $\Delta T_j(x, y, z)$ ($j = 1, 2, \ldots$) are defined as the j th Fourier coefficients of the perturbations of respectively the electron, hole and temperature distributions throughout this thesis. Though the use of half-amplitudes might look odd in formulas (3.1), (3.2) and (3.3), it proves necessary so as to comply with the definitions used in the rest of this thesis.

3.1 Refractive Index of Electrically Conductive Materials

Following Maxwell's wave equation [1], the complex refractive index of an electrically conductive material at the optical angular frequency ω_{probe} of the probe laser reads

$$\tilde{n} = n + ik = \sqrt{\tilde{\epsilon}} = \sqrt{\tilde{\epsilon}_{\text{lattice}}(\omega_{\text{probe}}) + \underbrace{\frac{i\sigma(\omega_{\text{probe}})}{\omega_{\text{probe}}\epsilon_0}}_{\tilde{\epsilon}_\sigma}}, \qquad (3.5)$$

where n and k are respectively the real and imaginary parts of the complex refractive index, also called respectively the (real) refractive index and the extinction coefficient. $\tilde{\epsilon}$ is the total dielectric constant, $\tilde{\epsilon}_{\text{lattice}}$ is the dielectric constant of the intrinsic semiconductor (no free carriers available for conduction), σ is the frequency-dependent electrical conductivity and ϵ_0 is the dielectric constant of vacuum. Equation (3.5) clearly highlights that there exist two contributions to the refractive index, (i) $\tilde{\epsilon}_{\text{lattice}}$ and (ii) $\tilde{\epsilon}_\sigma$. The first contribution, $\tilde{\epsilon}_{\text{lattice}}$, accounts for all band-to-band (or *inter-band*) effects. $\tilde{\epsilon}_{\text{lattice}}$ varies explicitly with temperature [2], electric field [3], and implicitly with free carrier concentration [via carrier-induced bandgap narrowing [4] (BGN) and band-filling [5] (BF)]. The second contribution, $\tilde{\epsilon}_\sigma$, contains the electrical conductivity (i.e. the free-carrier) information, and is linked to all *intra-band* effects. It varies explicitly with the free carrier concentration and implicitly with temperature (via the carrier mobility).

The previous considerations show that the perturbations of the refractive index can be of three kinds. First, a perturbation of the carrier (i.e. electron and/or hole) concentration would impact the refractive index via both $\tilde{\epsilon}_{\text{lattice}}$ and $\tilde{\epsilon}_\sigma$. Second, a perturbation of the temperature distribution would also modify the refractive index via $\tilde{\epsilon}_{\text{lattice}}$ and $\tilde{\epsilon}_\sigma$. Finally, the presence of an electric field would modify the refractive index via $\tilde{\epsilon}_{\text{lattice}}$. In other words and in all generality, a perturbation of the refractive index reads

$$\Delta\tilde{n} = \underbrace{\underbrace{\underbrace{\frac{\partial n}{\partial N}\Delta N + \frac{\partial n}{\partial P}\Delta P}_{\text{Drude+BGN+BF}} + \underbrace{\frac{\partial n}{\partial \vec{E}}.\Delta\vec{E}}_{\text{Pockels+Kerr}}}_{\text{electrorefraction}} + i\underbrace{\left(\underbrace{\frac{\partial k}{\partial N}\Delta N + \frac{\partial k}{\partial P}\Delta P}_{\text{Drude+BGN+BF}} + \underbrace{\frac{\partial k}{\partial \vec{E}}.\Delta\vec{E}}_{\text{Franz-Keldysh}}\right)}_{\text{electroabsorption}}}_{\text{electrooptical}}$$

$$+ \underbrace{\underbrace{\frac{\partial n}{\partial T}\Delta T}_{\text{thermorefraction}} + i\underbrace{\frac{\partial k}{\partial T}\Delta T}_{\text{thermoabsorption}}}_{\text{thermooptical}} . \qquad (3.6)$$

In the following Sections, we review the different effects and discuss their relative magnitudes for Si at a 670 nm wavelength (1.85 eV), i.e. for the probe laser of TP. Note, for the sake of completeness, that a very complete review of the published data concerning this problem can be found in Ref. [6] and, with even more details, in Ref. [7]. Unfortunately, numerically speaking, both authors limit the study to wavelengths relevant to fiber-optic communication ($\lambda = 1.3$ and $1.55\,\mu$m). These results can therefore not directly be used in our investigations. A conclusion which can be derived from these investigations, however, is that although all published models and experimental data agree qualitatively, very large variations in the derived numerical values of the derivatives of Eq. (3.6) are found (up to a factor 10!). In this Chapter, we therefore limit our study to the simplest models. Chapter 6 is anyhow dedicated to the comparison of the (complete) developed model with experimental data.

3.2 Electrooptical Effects

The electrooptical effects account for the changes in complex refractive index due to either the presence of free carriers or of an electric field. Three free-carrier electrooptical phenomena are to be reported, namely (i) the Drude effect (Sect. 3.2.1), (ii) the carrier-induced bandgap narrowing (BGN) effect (Sect. 3.2.2) and (iii) the band-filling (BF) effect (Sect. 3.2.3). In addition, three electric-field effects are to be taken into account, namely the (i) Kerr, (ii) Pockels and (iii) Franz-Keldysh effects (Sect. 3.2.4).

3.2.1 Drude Effect

The Drude effect accounts for both electrorefraction (i.e. change in n) and electroabsorption (i.e. change in k) due to a variation in electrical conductivity [1], i.e. a variation in $\tilde{\epsilon}_\sigma$. We show here that, at high optical frequencies like in red and NIR, the electrorefractive Drude effect is linear (Δn proportional to N) and the electroabsorptive Drude effect is negligible.

The frequency-dependent Drude electrical conductivity due to charge carriers of charge q and mobility μ and in concentration N_q is [8]

$$\sigma = \frac{q\mu N_q}{1 - i\omega_{\text{probe}}/\omega_{\text{scatt}}}, \tag{3.7}$$

where $\omega_{\text{scatt}} = q/(m\mu)$ is the scattering frequency of the carriers and m their mass. In silicon, this scattering frequency is much smaller [8] ($\sim$10 THz) than the optical frequencies ω_{probe} corresponding to red and NIR wavelengths ($\sim$1 PHz). In other words, the high-frequency electrical conductivity is purely imaginary (no resistive losses).

Using formula (3.5) and assuming a real $\tilde{\epsilon}_{\text{lattice}}$ (Ref. [9]), the complex refractive index then reads, using first-order Taylor expansion,

$$\tilde{n} = \sqrt{\tilde{\epsilon}_{\text{lattice}} + \frac{i}{\omega_{\text{probe}}\epsilon_0} \frac{q\mu N_q}{1 - i\omega_{\text{probe}}\frac{\mu m}{q}}} \approx \sqrt{\tilde{\epsilon}_{\text{lattice}} - \frac{q^2}{\omega_{\text{probe}}^2 \epsilon_0 m} N_q}$$
$$\approx n_0 \left(1 - \frac{q^2}{2\omega_{\text{probe}}^2 \epsilon_0 n_0^2 m} N_q\right). \tag{3.8}$$

The first derivatives of the real and imaginary refractive indices can therefore be written respectively

$$\frac{\partial n}{\partial N}|_{\text{Drude}} = -\underbrace{\frac{q^2}{2\omega_{\text{probe}}^2 \epsilon_0 n_0}}_{=\beta} \frac{1}{m_e} = -2.036 \times 10^{-22}\ \text{cm}^3 \tag{3.9a}$$

$$\frac{\partial n}{\partial P}|_{\text{Drude}} = -\underbrace{\frac{q^2}{2\omega_{\text{probe}}^2 \epsilon_0 n_0}}_{=\beta} \frac{1}{m_h} = -1.371 \times 10^{-22}\ \text{cm}^3 \tag{3.9b}$$

$$\frac{\partial k}{\partial N}|_{\text{Drude}} \approx \frac{\partial k}{\partial P}|_{\text{Drude}} \gtrsim 0, \tag{3.9c}$$

where β is the so-called Drude coefficient and m_e and m_h are respectively the electron and hole effective masses. The numerical values are given for the probe laser of TP (670 nm). Interestingly, the expression of β shows its independence from mobility and, hence, from temperature. The variations in $\tilde{\epsilon}_\sigma$ with temperature are therefore negligible in the considered wavelength range.

For completeness, notice that Eq. (3.9) are strictly only valid in visible and NIR wavelengths. Deeper in the infra-red, the extinction coefficient of Si strongly increases with free carrier concentration. This is often referred to as free carrier absorption (FCA) [8]. Notice also that, besides being very small in visible and NIR wavelengths, $\partial k/\partial N|_{\text{Drude}}$ and $\partial k/\partial P|_{\text{Drude}}$ are also always positive. In other words, the Drude contributions to the real and imaginary refractive indices have opposite signs and different orders of magnitude.

3.2.2 Carrier-Induced Bandgap Narrowing (BGN) Effect

The presence of free carriers in a silicon sample modifies the band structure of the lattice (Fig. 3.1a). The free carriers renormalize the sample bandgap E_g (Ref [4]), which, in turn, modifies the complex refractive index via $\tilde{\epsilon}_{\text{lattice}}$.

Physically, the change in band-to-band absorption coefficient α_{BTB} stems from the increase in available states for electron-hole pair generation for a fixed wavelength.

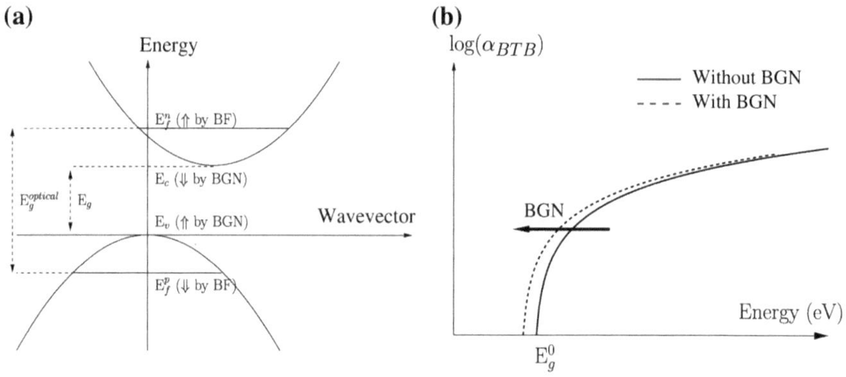

Fig. 3.1 **a** Schematic of the band structure of Si. The impact of BGN and BF effects are indicated between brackets. E_c is the lowest energy level of the conduction band and E_v the highest energy level of the valence band, the bandgap is $E_g = E_c - E_v$. E_f^n and E_f^p are respectively the electron and hole quasi-Fermi levels, the optical bandgap is $E_g^{\text{optical}} \approx E_f^n - E_f^p$. It is implicitly assumed in this figure that the semiconductor is highly degenerated (E_f^n and E_f^p are located in the bands). **b** Impact of BGN on the variation in α_{BTB} with photon energy. E_g^0 is the bandgap energy before BGN

Above the indirect absorption edge E_g, the band-to-band absorption coefficient is such that [8]

$$\alpha_{BTB} \propto (\hbar\omega_{\text{probe}} - E_g)^2, \tag{3.10}$$

where $\hbar = h/(2\pi)$, h being Planck constant. A narrowed bandgap induces therefore an increased absorption coefficient (see Fig. 3.1b), i.e. an increased extinction coefficient. Given the Kramers-Kronig (KK) relationships [10] between n and k, it is easy to understand that BGN also impacts n.

Though this contribution has been measured and studied in the literature [3, 11, 12], no model exists to quantify it, even approximately. To explain mathematically the BGN effect e.g. due to the presence of free electrons, the following equation should be used

$$\frac{\partial \tilde{n}}{\partial N}|_{\text{BGN}} = \frac{\partial E_g}{\partial N}\left(\frac{\partial n}{\partial E_g} + i\frac{\partial k}{\partial E_g}\right). \tag{3.11}$$

Two terms therefore need to be quantified. Attempts to quantify the first term, $\partial E_g/\partial N$, can be found in the literature [4]. The second term, $(\partial n/\partial E_g + i\partial k/\partial E_g)$, has, to the best of our knowledge, not been fully modeled. Since a simplistic theoretical approach for modeling this term proves to overestimate the impact of BGN on n and k [13], we believe the complexity of this issue falls out of the scope of our investigations.

As an alternative, we propose an approximate quantification technique based on the comparison of experimental spectra measured respectively on doped and undoped silicon. This study further needs the following three assumptions. First, though doping not only impacts E_g via the added free carriers (*plasma*- or *carrier*-induced BGN) but also through the presence of dopant impurities (*doping*-induced BGN), we neglect the latter. Second, we assume a linear BGN electrooptical effect, i.e. the n and k variations due to BGN are proportional to the free carrier concentration. Finally, we assume that the effect of BGN on the optical functions is independent from whether the BGN is induced by electrons or holes. None of these three assumptions is actually valid but they provide the framework for an approximate quantification of the BGN electrooptical effect. We believe that the obtained order of magnitude of the effect is acceptable.

We focus our study on the experimental data of Ref. [9] and Ref. [14], where broad spectra of n and k with a high density of data points are provided for undoped and highly doped ($N_{\text{doping}} \approx 3 \times 10^{20}\,\text{cm}^{-3}$) silicon. This is ideal for KK transform needed to study the electrorefractive BGN effect (see below). We compare in Fig. 3.2a–d the experimental spectra of respectively the real part of $\tilde{\epsilon}$, i.e. $\epsilon_1 = \Re(\tilde{\epsilon})$, the imaginary part of $\tilde{\epsilon}$, i.e. $\epsilon_2 = \Im(\tilde{\epsilon})$, n and k measured on lowly doped silicon and highly doped silicon. The observed difference between the spectra at low and high doping is due to the difference in free carrier concentrations, i.e. both to the Drude and BGN effects.[2]

Figure 3.2e shows the spectra of the differences in optical functions at high and low doping. Three quantities are to be discussed, i.e. the carrier-induced increase in extinction coefficient $|\Delta k|$, the total carrier-induced reduction in refractive index $|\Delta n_{\text{Tot}}|$ and the BGN-induced increase in refractive index $|\Delta n_{\text{BGN}}|$. First, it can be seen that, though small, the variation Δk is not negligible. As mentioned earlier, the Drude effect expects a negligible variation in k (at least an order of magnitude smaller than the observed variation). In other words, the observed $|\Delta k|$ is a BGN effect. In the particular case of the TP probe laser, we have $\Delta k = 0.00897$. Second, the total variation Δn_{Tot} in refractive index, resulting from a combination of the Drude and BGN effects, can also be observed in Fig. 3.2e. Interestingly, these variations are in good agreement with the Drude effect of Eq. (3.9a) in the energy below approximately 3 eV. This indicates a dominant Drude effect. Finally, Fig. 3.2e also shows the increase in n due to BGN. Though less direct, the Drude and BGN contributions to the total variation in refractive index can indeed easily be separated. The ϵ_2 spectra being free of any Drude contribution (in the range of interest, i.e. the red and NIR range), they can be KK transformed using the following formula

$$\epsilon_1^{\text{KK}}(E) = KK(\epsilon_2) = 1 + \frac{2}{\pi}\mathcal{P}\int_0^{+\infty} \frac{E'\epsilon_2(E')}{E'^2 - E^2}dE', \tag{3.12}$$

[2] The BF effect studied below will prove to be negligible

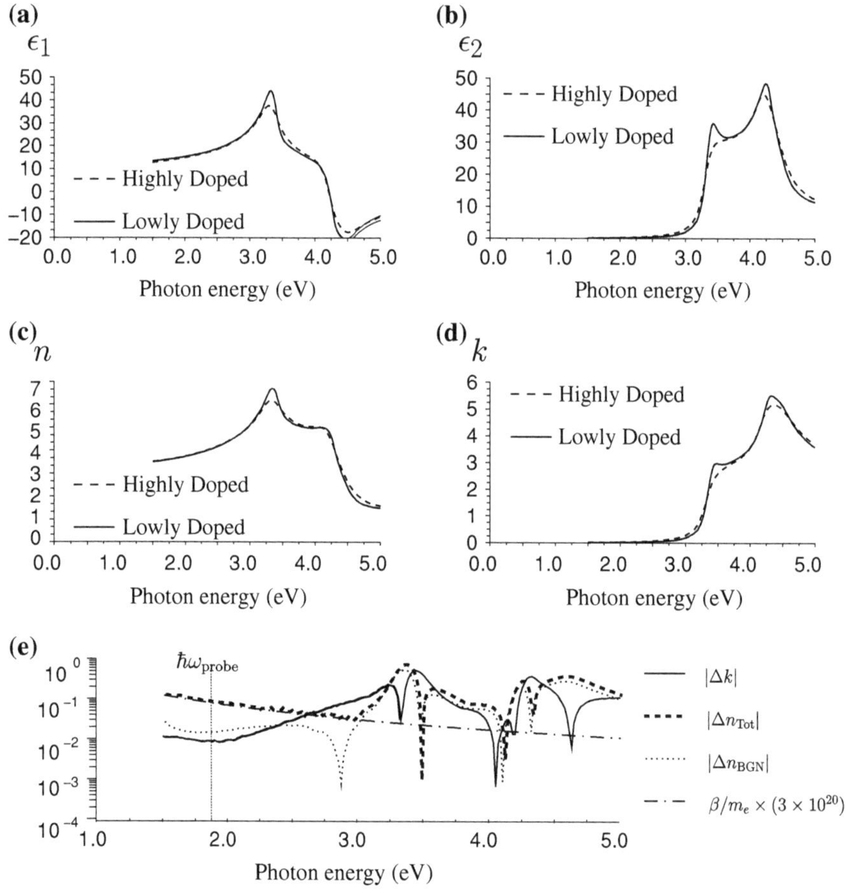

Fig. 3.2 **a** Experimental ϵ_1 spectra of respectively lowly doped (*full line*) and highly doped (*dashed line*) silicon. **b** Experimental ϵ_2 spectra of respectively lowly doped (*full line*) and highly doped (*dashed line*) silicon. **c** Experimental n spectra of respectively lowly doped (*full line*) and highly doped (*dashed line*) silicon. **d** Experimental k spectra of respectively lowly doped (*full line*) and highly doped (*dashed line*) silicon. **e** Spectra of the differences $|\Delta k|$ (*full line*) and $|\Delta n|$ (*dashed line*) measured respectively on highly doped and lowly doped silicon. The dotted line represents the spectrum of the BGN-induced variation Δn_{BGN} in n as obtained from KK transform. The interrupted line shows the theoretical electrorefractive Drude effect (Eq. (3.9a)). The vertical *dotted line* indicates the energy of the TP probe laser (1.85 eV)

where $\mathcal{P}$ is the principal value of the integral and E is the photon energy [10]. The obtained ϵ_1^{KK} spectra are then equally free of any Drude contribution. The difference in the resulting n spectra (i.e. $\Re(\sqrt{\epsilon_1^{\mathrm{KK}} + \epsilon_2})$) gives therefore the BGN-induced variation Δn_{BGN} in real refractive index. It can be seen that, around the energy of the TP probe laser, though larger than Δk, Δn_{BGN} is slightly less than one order of magnitude smaller than the total variation in n. In the particular case of the TP probe laser, we obtain $\Delta n_{\mathrm{Tot}} = -0.0654$ and $\Delta n_{\mathrm{BGN}} = 0.0146$.

Combining these results with the three assumptions mentioned above, we derive

$$\frac{\partial n}{\partial N}|_{\mathrm{BGN}} = \frac{\partial n}{\partial P}|_{\mathrm{BGN}} = 5 \times 10^{-23}\ \mathrm{cm}^3 \tag{3.13a}$$

$$\frac{\partial k}{\partial N}|_{\mathrm{BGN}} = \frac{\partial k}{\partial P}|_{\mathrm{BGN}} = 3 \times 10^{-23}\ \mathrm{cm}^3, \tag{3.13b}$$

for the probe laser of TP.

Hence, following this approach, the electrorefractive BGN effect is smaller than the electrorefractive Drude effect (Eqs. (3.9a) and (3.9b)). It further has opposite sign, reducing the total impact of the carrier upon the real refractive index. We will see in Chap. 6, however, that a reduced impact of the free carriers is to be rejected. Based on our experimental data, we will indeed conclude that the electrorefractive effect is underestimated. For this reason, we neglect this contribution in the rest of this thesis.

The electroabsorptive BGN effect has no Drude counterpart and should therefore be included in our theory. However, it is about one order of magnitude smaller than the Drude electrorefractive effect and can consequently be considered as a second-order effect. Though we do not include this effect in our main theory, we still look for traces of this effect in Appendix A.6. The final conclusions of these investigations are that it is never observed on ultra-shallow junctions.

3.2.3 Burstein Shift or Band-Filling (BF) Effect

Similar to the BGN effect, when free carriers are present in a silicon sample, the optical bandgap also suffers from modifications due to BF (Fig. 3.1a). Free electrons indeed fill the energy states at the bottom of the conduction band, making them no longer available for absorption by a valence electron. Free holes empty the energy states at the top of the valence band, leading to similar reduction in absorption coefficient. In summary, free carriers widen the optical bandgap, which in turn diminishes the absorption coefficient.

However, we do not expect any Burstein shift in our experiments. This effect is indeed only effective when a carrier density is reached such that the carrier distributions are degenerate. It is usually assumed that this occurs when the electron (resp. hole) quasi-Fermi level lies $4k_bT$ above (resp. below) the bottom of the conduction band (resp. the top of the valence band), k_b being Boltzmann's constant and T the lattice temperature. In Si at 300 K, this corresponds respectively to an electron concentration of $1.7 \times 10^{20}\ \mathrm{cm}^{-3}$ and a hole concentration of $6.4 \times 10^{19}\ \mathrm{cm}^{-3}$. These concentrations are hardly ever reached in our experiments. Besides, even in cases of higher carrier densities, only *enhanced* absorption is observed [3, 11, 12, 15], i.e. BGN-induced variations in refractive index seem to dominate in silicon. This is confirmed by the experimental spectra shown in Fig. 3.2. This effect is therefore neglected in this work. It is, however, important to keep this effect in mind in the case of small effective mass semiconductors (such as InSb), where this effect can be considerable [5].

3.2.4 Pockels, Kerr and Franz-Keldysh Effects

The Pockels and Kerr effects are respectively first- and second-order electrorefractive effects due to the presence of an electric field [3, 8]. The Franz-Keldysh effect accounts for electroabsorption due to the presence of an electric field [3, 16] (enhanced band-to-band absorption under high electric field due to band bending). These effects impact $\tilde{\epsilon}_{\text{lattice}}$. A very thorough investigation of these phenomena has been conducted by Aspnes [17]. These effects can be taken advantage of in order to quantify the electric field at the junction between two semiconductors [18] or at the surface of a semiconductor sample [19].

In the case of TP, these three effects can be neglected for two reasons. First, they are significant only at wavelengths very close to the (direct or indirect) bandgap of the studied semiconductor [3, 20]. Second, for reflection to be sensitive to an electric field, the in-depth extension of the electric field needs to be of the order of the wavelength. The Debye length, giving the screening length of an electric field by free carriers in density N_q is [21]

$$L_{\text{Debye}} = \sqrt{(\epsilon_{\text{lattice}}\epsilon_0 k_b T)/(q^2 N_q)}, \tag{3.14}$$

where k_b is Boltzmann constant. In a TP experiment, the involved carrier concentrations are at least $\approx 10^{18}\,\text{cm}^{-3}$, the extension of any electric field is therefore only a couple of nanometers. In summary, due to the high carrier concentrations present in the sample, an electric field $\vec{E}(x, y, z, t)$ generates a very local peak of refractive index perturbation, which induces negligible perturbation of the reflectance. The Pockels, Kerr and Franz-Keldysh effects are therefore not considered in the rest of this thesis.

3.3 Thermooptical Effects

In all generality, a variation in sample temperature modifies the complex refractive index via both $\tilde{\epsilon}_{\text{lattice}}$ and $\tilde{\epsilon}_\sigma$. However, the variations in $\tilde{\epsilon}_\sigma$ have been demonstrated to be negligible in the considered wavelength range (Sect. 3.2.1). The temperature therefore only impacts the complex refractive index through $\tilde{\epsilon}_{\text{lattice}}$. These variations stem mostly from the thermally induced BGN and partly from thermal expansion [22].

To model the total thermooptical effects, we use a fitting of experimental data [2]. This fitting shows that the thermoabsorption is negligible with respect to thermorefraction and that thermorefractive effect is linear, i.e.

$$\frac{\partial \tilde{n}}{\partial T} = \delta = 3.71 \times 10^{-4}\,\mathrm{K}^{-1}, \tag{3.15}$$

where the value is given for the probe laser of TP.

Very importantly for the rest of this work, we would like to underline the fact that the thermorefractive and electrorefractive effects have opposite signs (Eqs. (3.9) vs (3.15)), i.e. an increased carrier concentration reduces the real refractive index while an increased temperature raises the real refractive index. This competing behavior proves to be dominated by the electrorefractive effect in most cases studied in this work. Using Eqs. (3.9) and (3.15), it is easy to demonstrate that the thermorefractive effect indeed only becomes dominant when $\Delta T(\mathrm{K}) \gtrsim 10^{-18} \Delta N(\mathrm{cm}^{-3})$ if we assume equal free electron and hole concentrations (ΔN). We will see in Chap. 6 that the latter situation is never observed on the DC reflectance (free carriers due to active doping always dominate). We will, however, see that the thermorefractive effect dominates the AC reflectance in the case of highly doped substrates (active doping $\gtrsim 5 \times 10^{19}\,\mathrm{cm}^{-3}$). This will lead to a very interesting behavior of the AC reflectance when the electrorefractive and thermorefractive effects are commensurate (active doping $\approx 10^{19}\,\mathrm{cm}^{-3}$). Similarly, the AC reflectance on highly damaged samples, e.g. implantation profiles before annealing, is also dominated by the thermorefractive effect [23–31]. This last case is, however, not the focus of this work.

3.4 Summary

In summary, for TP, the electrooptical and thermooptical effects in silicon can be summarized by adding formulas (3.9) and (3.15), i.e.

$$\Delta \tilde{n} = \Delta n = \frac{\partial n}{\partial \Delta N}\Delta N + \frac{\partial n}{\partial \Delta P}\Delta P + \frac{\partial n}{\partial T}\Delta T = -\beta \frac{1}{m_e}\Delta N - \beta \frac{1}{m_h}\Delta P + \delta \Delta T. \tag{3.16}$$

As highlighted by Eq. (3.16), only real variations in refractive index are expected. Furthermore, the variations due to an increase in free carrier concentration or temperature have opposite signs.

Formula (3.16) is in agreement with the commonly used optical model for PMOR [32, 33]. It has, however, to be kept in mind that formula (3.16) assumes negligible impact of BGN on the complex refractive index. This is discussed in more detail in Appendix A.6 where we look for traces of the BGN effect in our experimental data.

References

1. M. Born, E. sored Wolf, Principles of Optics (Pergamon, Oxford, UK, 1970).
2. G.E. Jellison, F.A. Modine, Optical functions of silicon at elevated temperatures. J. Appl. Phys. **76**(6), 3758–3761 (1994)
3. R.A. Soref, B.R. Bennett, Electrooptical effects in silicon. IEEE J. Quantum Electron. **23**(1), 123–129 (January 1987)
4. A. Schenk, Advanced Physical Models for Silicon Device Simulation (Springer, Vienna, 1998)
5. E. Burstein, Anomalous optical absorption limit in insb. Phys. Rev. **93**(3), 632–633 (1954)
6. L. Pavesi, G. Guillot, *Optical Interconnects. The silicon Approach* (Springer, Berlin, 2006)
7. R. Thalhammer, Internal Laser Probing Techniques for Power Devices: Analysis, Modeling and Simulation. PhD thesis, Universitat Muchen, 2000, URL http://tumb1.biblio.tu-muenchen.de/publ/diss/ei/2000/thalhammer.pdf
8. M. Fox, *Optical Properties of Solids* (Oxford University Press, Oxford, 2001)
9. T. Yasuda, D.E. Aspnes, Optical-standard surfaces of single-crystal silicon for calibrating ellipsometers and reflectometers. Appl. Opt. **33**(31), 7435–7438 (1994)
10. K.E. Peiponen Lucarini, J.J. Saarinen, E.M. Vartiainen, Kramers-Kronig relations in optical materials research. Springer Series, 2005
11. P.E. Schmid, Optical absorption in heavily doped silicon. Phys. Rev. B **23**(10), 5531–5536 (1981), URL http://link.aps.org/abstract/PRB/v23/p5531
12. G.E. Jellison, S.P. Withrow, J.W. Mccamy, J.D. Budai, D. Lubben, M.J. Godbole, Optical functions of ion-implanted, laser-annealed heavily-doped silicon. Phys. Rev. B **52**(20), 14607–14614 (1995)
13. J. Bogdanowicz, F. Dortu, T. Clarysse, W. Vandervorst, A. Salnik, Electrothermal theory of photomodulated optical reflectance on active doping profiles in silicon. J. Appl. Phys. **108**(10), 104908 (2010)
14. D.E. Aspnes, A.A. Studna, E. Kinsbron, Dielectric-properties of heavily doped crystalline and amorphous-silicon from 1.5 to 6.0 ev. Phys. Rev. B **29**(2), 768–779 (1984)
15. H.C. Huang, S. Yee, M. Soma. Quantum calculations of the change of refractive-index due to free-carriers in silicon with nonparabolic band-structure. J. Appl. Phys. **67**(4), 2033–2039 (1990)
16. P.H. Wenland, M. Chester, Electric field effects on indirect optical transitions in silicon. Phys. Rev. **140**, A1384–A1390 (1965)
17. D.E. Aspnes, *Handbook on Semiconductors: Modulation Spectroscopy* (T.S, Moss, 1980)
18. C. Vanhoof, K. Deneffe, J. Deboack, D.J. Arent, G. Borghs, Franz-keldysh oscillations originating from a well-controlled electric-field in the gaas depletion region. Appl. Phys. Lett. **54**(7), 608–610 (February 1989)
19. T. Zhang, M. Di, E.J. Bersch, H. Chouaib, A. Salnik, L. Nicolaides, S. Consiglio, R.D. Clark, A.C. Diebold, Photoreflectance spectroscopic characterization of si with sio2 and hfo2 dielectric layers. in *Proceedings of the International Conference on Frontiers of Characterization and Metrology for Nanoelectronics*, 2009
20. B.O. Seraphin, N. Bottka, Band-structure analysis from electro-reflectance studies. Phys. Rev. **145**(2), 628–636 (May 1966). doi:10.1103/PhysRev.145.628
21. B.G. Streetman, S. Banerjee, *Solid State Electronic Devices*, 5th edn. (Prentice Hall International, Inc., Englewood Cliffs, 2000)
22. G. Ghosh, Temperature dispersion of refractive-indexes in crystalline and amorphous-silicon. Appl. Phys. Lett. **66**(26), 3570–3572 (June 1995)
23. L. Nicolaides, A. Salnick, J. Opsal, Study of low energy implants for ultrashallow junctions using thermal wave and optical techniques. Rev. Sci. Instrum. **74**(1), 563–565 (2003)
24. L. Nicolaides, A. Salnick, J. Opsal, Nondestructive analysis of ultrashallow junctions using thermal wave technology. Rev. Sci. Instrum. **74**(1), 586–588 (2003)
25. A. Salnick, J. Opsal, Quantitative photothermal characterization of ion-implanted layers in si. J. Appl. Phys. **91**(5), 2874–2882 (2002)

26. G. Smets, E. Rosseel, G. Sterckx, J. Bogdanowicz, W. Vandervorst, D. Shaughnessy, Transfer from rs-based to pmor-based ion implantation process monitoring. AIP Conf. Proc. **1321**, 426–31 (2011)
27. P. Borden, C. Ferguson, D. Sing, L. Larson, L. Bechtler, K. Jones, P Gable, In-line characterization of preamorphous implants (pai). Technical report, BoxerCross, 2000
28. C. Christofides, I.A. Vitkin, A. Mandelis, Photothermal reflectance investigation of processed silicon. I. Room-temperature study of the induced damage and of the annealing kinetics of defects in ion-implanted wafers. J. Appl. Phys. **67**(6), 2815–21 (March 1990)
29. B.C. Forget, D. Fournier, Characterization of implanted silicon-wafers by the nonlinear photoreflectance technique. Mater. Sci. Eng. B-Solid State Mater. Adv. Technol. **24**(1–3), 199–202 (1994)
30. B.C. Forget, D. Fournier, V.E. Gusev, Non-linear recombination processesapplication to quantitative implantation characterization. J. Phys. IV **4**(C7), 155–158 (1994)
31. W. Vandervorst, T. Clarysse, B. Brijs, R. Loo, Y. Peytier, B. J. Pawlak, E. Budiarto, P. Borden, Carrier illumination as a tool to probe implant dose and electrical activation. in International Conference on Characterization and Metrology for ULSI Technology, ed. by D.G. Seiler, A.C. Diebold, T.J. Shaffner, R. McDonald, S. Zollner, R.P. Khosla, E.M. Secula. vol. 683 (AIP), pp. 758–763, 2003, URL http://link.aip.org/link/?APC/683/758/1
32. F. Dortu, T. Clarysse, R. Loo, B. Pawlak, R. Delhougne, W. Vandervorst, Progress in the physical modeling of carrier illumination. J. Vac. Sci. Technol. B **24**(3), 1131–1138 (2006)
33. R.E. Wagner, A. Mandelis, A generalized calculation of the temperature and drude photo-modulated optical reflectance coefficients in semiconductors. J. Phys. Chem. Solids **52**(9), 1061–1070 (1991)

Chapter 4
Theory of Carrier and Heat Transport in Homogeneously Doped Silicon

In this chapter, we assume that a supra-bandgap modulated pump laser and a supra-bandgap constant probe laser shine on a homogeneously doped silicon sample with respective irradiances $\Pi_{\text{pump}}(x, y, t)$ and $\Pi_{\text{probe}}(x, y)$ such that

$$\begin{aligned}\Pi_{\text{pump}}(x, y, t) &= \mathcal{P}_{\text{pump}}(t) \exp\left(-\frac{x^2 + y^2}{\mathcal{R}^2_{\text{pump}}}\right) \\ &= \mathcal{P}^0_{\text{pump}} \exp\left(-\frac{x^2 + y^2}{\mathcal{R}^2_{\text{pump}}}\right)\left[\frac{1}{2} + \frac{1}{2}\cos(\omega t)\right] \\ &= \mathcal{P}^0_{\text{pump}} \exp\left(-\frac{x^2 + y^2}{\mathcal{R}^2_{\text{pump}}}\right)\left[\frac{1}{2} + \frac{1}{4}\exp(i\omega t) + \frac{1}{4}\exp(-i\omega t)\right]\end{aligned} \tag{4.1}$$

$$\Pi_{\text{probe}}(x, y) = \mathcal{P}^0_{\text{probe}} \exp\left(-\frac{x^2 + y^2}{\mathcal{R}^2_{\text{probe}}}\right), \tag{4.2}$$

where $\mathcal{P}_{\text{pump}}(t)$ is the time-dependent pump irradiance at position $(x, y) = (0, 0)$, $\mathcal{P}^0_{\text{pump}}$ (resp. $\mathcal{P}^0_{\text{probe}}$) is the peak value of the pump (resp. probe) irradiance, $\mathcal{R}_{\text{pump}}$ (resp. $\mathcal{R}_{\text{probe}}$) is the pump (resp. probe) radius and ω is the angular modulation frequency of the pump laser irradiance.

As a result, a free electron distribution $N(x, y, z, t)$, a free hole distribution $P(x, y, z, t)$ and a temperature distribution $T(x, y, z, t)$ are generated in the sample such that

$$\begin{aligned}N(x, y, z, t) &= N_{\text{doping}} + \Delta N(x, y, z, t) \\ &= N_{\text{doping}} + \underbrace{\Delta N_0(x, y, z)}_{\text{dc}} \\ &\quad + \underbrace{\left[\Delta N_1(x, y, z)\exp(i\omega t) + \Delta N_1^\star(x, y, z)\exp(-i\omega t)\right]}_{\text{fundamental}}\end{aligned}$$

J. Bogdanowicz, *Photomodulated Optical Reflectance*, Springer Theses,
DOI: 10.1007/978-3-642-30108-7_4,

$$+\sum_{j=2}^{+\infty}\underbrace{\left[\Delta N_j(x,y,z)\exp(ij\omega t)+\Delta N_j^{\star}(x,y,z)\exp(-ij\omega t)\right]}_{j^{\text{th}}\text{ harmonic}} \tag{4.3}$$

$$\begin{aligned} P(x,y,z,t) &= P_{\text{doping}}+\Delta P(x,y,z,t)\\ &= P_{\text{doping}}+\underbrace{\Delta P_0(x,y,z)}_{\text{dc}}\\ &\quad+\underbrace{\left[\Delta P_1(x,y,z)\exp(i\omega t)+\Delta P_1^{\star}(x,y,z)\exp(-i\omega t)\right]}_{\text{fundamental}}\\ &\quad+\sum_{j=2}^{+\infty}\underbrace{\left[\Delta P_j(x,y,z)\exp(ij\omega t)+\Delta P_j^{\star}(x,y,z)\exp(-ij\omega t)\right]}_{j^{\text{th}}\text{ harmonic}} \end{aligned} \tag{4.4}$$

$$\begin{aligned} T(x,y,z,t) &= T_0+\underbrace{\Delta T_0(x,y,z)}_{\text{dc}}\\ &\quad+\underbrace{\left[\Delta T_1(x,y,z)\exp(i\omega t)+\Delta T_1^{\star}(x,y,z)\exp(-i\omega t)\right]}_{\text{fundamental}}\\ &\quad+\sum_{j=2}^{+\infty}\underbrace{\left[\Delta T_j(x,y,z)\exp(ij\omega t)+\Delta T_j^{\star}(x,y,z)\exp(-ij\omega t)\right]}_{j^{\text{th}}\text{ harmonic}}, \end{aligned} \tag{4.5}$$

where N_{doping} (resp. P_{doping}) and $\Delta N(x,y,z,t)$ (resp. $\Delta P(x,y,z,t)$) are respectively the electron (resp. hole) concentrations due to the homogeneous doping of the sample and due to optical injection. T_0 is the environment temperature during the experiment, i.e. room temperature (300 K). $\Delta N_0(x,y,z)$, $\Delta P_0(x,y,z)$ and $\Delta T_0(x,y,z)$ are the time-independent components of the laser-induced perturbations respectively of the free electron distribution, of the free hole distribution and of the temperature distribution. $\Delta N_1(x,y,z)$, $\Delta P_1(x,y,z)$ and $\Delta T_1(x,y,z)$ are the fundamental modes of the laser-induced perturbations respectively of the free electron distribution, of the free hole distribution and of the temperature distribution. $\Delta N_j(x,y,z)$, $\Delta P_j(x,y,z)$ and $\Delta T_j(x,y,z)$ are the jth Fourier coefficients of the laser-induced perturbations respectively of the free electron distribution, of the free hole distribution and of the temperature distribution. $\Delta N_j^{\star}$, $\Delta P_j^{\star}$ and $\Delta T_j^{\star}$ ($j=1,2,\ldots$) are the complex conjugates of jth Fourier coefficients of respectively the electron, hole and temperature distributions.

At this point, we believe it is a good opportunity to make the following remark concerning the use of complex notations in this thesis. Both the *optical* model and the *transport* model involve a modulation and, hence, benefit from the convenient use of a complex notation. First, due to the modulation of the electromagnetic field at the *optical frequency of the probe laser*, the optical model uses a complex refractive index (Eq. 3.5). Second, the transport model includes the *modulation of the pump irradiance* (Eq. (4.1)). The pump-generated excess carriers and temperature therefore also are written as complex numbers (Eqs. (4.3)–(4.5)). Obviously, these two complex notations should not be confused or intermixed. Luckily, the ambiguity can be eliminated. Indeed, the final equations of the optical model (formulas (2.11), (2.17), (2.25), (2.31) and (3.16)) are purely real. As of this chapter, the only complex notation is therefore related to the time variations in the excess carrier concentration and temperature at the modulation frequency of the pump power. This explains why the free electron, free hole and temperature distributions are written as real numbers in Chap. 3 and as complex numbers in the present chapter. The two notations are obviously equivalent.[1]

In this chapter, we would like to calculate the various components of the free electron, free hole and temperature distributions (Eqs. (4.3)–(4.5)) generated by the pump and probe irradiances (Eqs. (4.1) and (4.2)). In particular, in Sect. 4.1, we introduce the thermodynamic model of carrier and heat transport. In Sect. 4.1.1, we show that it is possible to simplify the carrier transport equations into one single equation, namely the generalized ambipolar diffusion equation for the carrier transport. Similarly, in Sect. 4.1.2, we simplify the heat transport problem. In Sect. 4.1.3, we then study the equations which need to be solved to derive the various harmonics of the electron, hole and temperature distributions. Unlike in the previous chapters, the here considered perturbations of the electron and hole concentrations can be large compared to their equilibrium values. We show that this implies frequency mixing, i.e. coupled equations. An in-depth discussion of the time-dependent transport equations is therefore needed. Section 4.2 is then dedicated to the study and discussion of the solutions of the derived carrier and heat transport equations.

Note that, though this chapter focuses on homogeneously doped Si, the derived results and understanding are of the highest importance for the next chapter, where the carrier and heat transport theory is studied in non-homogeneously doped Si. Chapter 5 will indeed show that, when a doped layer is present at the surface of a silicon sample, the excess carrier and temperature distributions can easily be derived from the distributions found in a homogeneously doped sample, as studied in the present chapter.

4.1 Thermodynamic Model

The thermodynamic model is an extension of the drift-diffusion model including the thermoelectric effects, i.e. the interactions between temperature and carriers. These effects have hardly ever been discussed in the framework of PMOR. Wagner [1] incorporated these effects *a posteriori* in his model to show that they should be negligible.

[1] The required equivalence justifies the use of the half-amplitudes in Eqs. (3.1)–(3.3)

Opsal [2] also studied the effect of thermodiffusion of carriers (similar to a Seebeck effect) but only considered constant thermodiffusion coefficients. We, however, take these thermoelectric effects into account *a priori* and then show consistently when they are negligible. We also include the bandgap narrowing quasi-drift.

In this work, we make use of the model introduced by Kells (Sect. 2.3 of Ref. [3]). More particularly, we use the simplified model derived in Sect. 2.3.6 of the same reference [3], where it is assumed that electrons and holes are in thermal equilibrium with the lattice (electron temperature T_n, hole temperature T_p and lattice temperature T are equal). This assumption is acceptable for silicon if the investigated time scale is not shorter than a few picoseconds, i.e. the thermalization time of hot carriers [4].

In the thermodynamic model, the *carrier transport equations*, i.e. Poisson's equation and the electron and hole continuity equations, are written in their usual form as

$$-\vec{\nabla}\epsilon_0\epsilon_{\text{lattice}}.\vec{\nabla}\psi = q(P - N + N_d^+ - N_a^-) \tag{4.6a}$$

$$\frac{\partial N}{\partial t} = \frac{1}{q}\vec{\nabla}.\vec{J}_n + G - \text{Rec} \tag{4.6b}$$

$$\frac{\partial P}{\partial t} = -\frac{1}{q}\vec{\nabla}.\vec{J}_p + G - \text{Rec}, \tag{4.6c}$$

where ψ is the electrostatic potential linked to the total electric field $\vec{E}$ by the following relationship

$$\vec{E} = -\vec{\nabla}\psi. \tag{4.7}$$

N_d^+ and N_a^- are respectively the ionized donor and acceptor concentrations. G is the total carrier generation term, Rec is the carrier recombination rate. $\vec{J}_n$ and $\vec{J}_p$ are respectively the electron and hole current densities. In our model, these currents respectively read

$$\vec{J}_n = \underbrace{-q\mu_n N\vec{\nabla}\psi}_{\vec{J}_n^{\text{drift}}} + \underbrace{qD_n\vec{\nabla}N}_{\vec{J}_n^{\text{diffusion}}} + \underbrace{k_b\mu_n N\vec{\nabla}T}_{\vec{J}_n^{\text{Seebeck}}} - \underbrace{\mu_n N\vec{\nabla}\chi}_{\vec{J}_n^{\text{BGN}}} \tag{4.8a}$$

$$\vec{J}_p = \underbrace{-q\mu_p P\vec{\nabla}\psi}_{\vec{J}_p^{\text{drift}}} - \underbrace{qD_p\vec{\nabla}P}_{\vec{J}_p^{\text{diffusion}}} - \underbrace{k_b\mu_p P\vec{\nabla}T}_{\vec{J}_p^{\text{Seebeck}}} - \underbrace{\mu_p P\vec{\nabla}(\chi + E_g)}_{\vec{J}_p^{\text{BGN}}}, \tag{4.8b}$$

where μ_n and μ_p are the electron and hole mobilities respectively, D_n and D_p are respectively the electron and hole diffusivities and χ is the electron affinity of the considered semiconductor.

Both the electron and hole currents proposed in formulas (4.8a, 4.8b) contain four components. First, the usual *drift* contribution $\vec{J}_i^{\text{drift}}$ ($i = n$ or p) accounts for the movements of charge under an applied electric field (not considered here) and internal electric fields. Two examples of such internal fields are of importance in this work. The Dember electric field generated by moving distributions of charges with opposite signs is studied in Sect. 4.1.1. The built-in electric field of a diode is discussed in Chap. 5. The second current contribution is the *diffusion* component $\vec{J}_i^{\text{diffusion}}$ ($i = n$ or p), which accounts for the displacement of charges towards regions of low concentration. Third, the thermodynamic model adds a current term $\vec{J}_i^{\text{Seebeck}}$ ($i = n$ or p) proportional to the temperature gradient to embody the *Seebeck* effect. Finally, we have also included the *BGN quasi-drift* currents [5, 6] $\vec{J}_i^{\text{BGN}}$ to account for the drift of electrons under gradients of affinity and the drift of holes under gradients of both affinity and bandgap. It has indeed been shown that the BGN quasi-drift currents are needed to model PMOR correctly [7]. We show in Sect. 4.1.1 that this BGN-induced drift acts as a counter-diffusive term.

In the thermodynamic model, the *heat transport equation* reads

$$\begin{aligned}\rho c_p \frac{\partial T}{\partial t} &= k_{th}\nabla^2 T + \underbrace{(h\nu_{\text{pump}} - E_g)G}_{G_{th}^{\text{direct}}} + \underbrace{\left[E_f^n - E_f^p + qT(P_p - P_n)\right]\text{Rec}}_{G_{th}^{\text{recombination}}} \\ &\quad + \underbrace{\frac{|\vec{J}_n|^2}{q\mu_n N} + \frac{|\vec{J}_p|^2}{q\mu_p P}}_{G_{th}^{\text{Joule}}} \underbrace{- T\left(\vec{J}_n . \frac{\partial P_n}{\partial N}\vec{\nabla} N + \vec{J}_p . \frac{\partial P_p}{\partial P}\vec{\nabla} P\right)}_{G_{th}^{\text{Peltier}}} \\ &\quad \underbrace{- T\left(\vec{J}_n . \frac{\partial P_n}{\partial T}\vec{\nabla} T + \vec{J}_p . \frac{\partial P_p}{\partial T}\vec{\nabla} T\right)}_{G_{th}^{\text{Thomson}}}, \end{aligned} \tag{4.9}$$

where ρ is the density of Si, c_p its heat capacitance, k_{th} is its thermal conductivity. E_f^n and E_f^p are respectively the electron and hole quasi-Fermi levels. P_n and P_p are respectively the electron and hole thermoelectric powers (see Sect. 4.1.2).

The heat equation shows five heat generation terms. First, G_{th}^{direct} is due to the fast *thermalization of hot carriers*. Second, $G_{th}^{\text{recombination}}$ is the heat released by carrier *recombinations*. Third, the *Joule heating* G_{th}^{Joule} occurs whenever a current flows through any electrically resistive material. Finally the fourth and fifth contributions represent respectively the Peltier heating G_{th}^{Peltier}, arising whenever an electrical current crosses a gradient of thermoelectric power, and the Thomson heating G_{th}^{Thomson}, which takes place whenever an electrical current flows through a temperature gradient.

In the rest of this Section, we look at both the carrier and heat transport equations above and simplify them in the specific case of TP. This leads to the *generalized ambipolar diffusion equation* (Sect. 4.1.1) and a simplified *heat equation* (Sect. 4.1.2). In order to derive separate equations for the different Fourier coefficients of the carrier and temperature distributions of Eqs. (4.3)–(4.5), we proceed with an in-depth discussion of the time dependence of the coefficients appearing in the final equations in Sect. 4.1.3. This will lead to our final steady-periodic model equations, i.e. a system of four coupled nonlinear equations solving for $\Delta N_0(x, y, z)$ $[= \Delta P_0(x, y, z)]$, $\Delta N_1(x, y, z)$ $[= \Delta P_1(x, y, z)]$, $\Delta T_0(x, y, z)$ and $\Delta T_1(x, y, z)$.

4.1.1 Generalized Ambipolar Diffusion Equation

Rather than directly solving the heavy problem of the carrier transport Eqs. (4.6a–4.6a) combined with the current Eqs. (4.8a, 4.8b), we would like to simplify it to a single equation, namely the generalized ambipolar diffusion equation. To attain this objective, the following four steps are needed.

The first simplification is the *charge balance* approximation (electrical neutrality). This assumes that the laser-induced electron and hole distributions are equal everywhere and at all times [i.e. $\Delta N(x, y, z, t) = \Delta P(x, y, z, t)$]. This obviously assumes no trapping but, most of all, it also supposes that electrons and holes diffuse and drift at the same speed, in spite of their different mobilities and, hence, diffusivities [8]. This approximation is realistic due to the presence of the internal Dember electric field which is generated by separated electron and hole distributions [9]. As Poisson's Eq. (4.6a) shows, in a homogeneously doped silicon sample, any disparity $(\Delta P - \Delta N)$ creates an internal field such that

$$\vec{\nabla}.\vec{E} = \frac{q}{\epsilon_0 \epsilon_{\text{lattice}}}(\Delta P - \Delta N). \tag{4.10}$$

This field tends to slow electrons down and accelerate holes so as to keep their density equal everywhere.

Since the internal field is created by the difference $(\Delta P - \Delta N)$ itself, the electron and hole densities cannot be *exactly* the same, i.e. the charge balance is only an approximation. It is, however, a good approximation if the disparity $(\Delta P - \Delta N)$ needed to create the field is small compared to the actual excess carrier densities ΔN and ΔP. McKelvey showed that, in the isothermal case, the charge balance is a good approximation if the Debye length is much smaller than the carrier diffusion length (p. 331–332 of Ref. [9]), which is always the case in Si (nm vs. μm). In the non-isothermal case, one needs to make sure that the internal field is also strong enough to counteract the natural tendency of electrons and holes to separate via their Seebeck currents. We follow the same reasoning as McKelvey did in the isothermal case. Assuming charge balance in the current Eq. (4.17) presented below, we can write the total current density

$$\vec{J}_{\text{TOT}} = \vec{J}_n + \vec{J}_p = q(\mu_n N + \mu_p P)\vec{E} + q(D_n^{TOT} - D_p^{TOT})\vec{\nabla}(\Delta N) \\ + k_b(\mu_n N - \mu_p P)\vec{\nabla}T. \tag{4.11}$$

The internal field must be large enough so as to ensure that the total current density is zero (no applied field), i.e.

$$\vec{E} = -\frac{(D_n^{TOT} - D_p^{TOT})}{(\mu_n N + \mu_p P)}\vec{\nabla}(\Delta N) - \frac{k_b(\mu_n N - \mu_p P)}{q(\mu_n N + \mu_p P)}\vec{\nabla}T. \tag{4.12}$$

Plugging this electric field into Eq. (4.10) leads to

$$-\frac{(D_n^{TOT} - D_p^{TOT})}{(\mu_n N + \mu_p P)}\nabla^2(\Delta N) - \frac{k_b(\mu_n N - \mu_p P)}{q(\mu_n N + \mu_p P)}\nabla^2 T = \frac{q}{\epsilon_0\epsilon_{\text{lattice}}}(\Delta P - \Delta N). \tag{4.13}$$

For the sake of simplicity, the coefficients of the two gradients in the left hand-side of Eq. (4.13) have been considered uniform. The relative difference between the excess electron and hole concentrations can therefore be written

$$\left|\frac{\Delta P - \Delta N}{\Delta N}\right| = \left|\frac{\epsilon_0\epsilon_{\text{lattice}}(D_n^{TOT} - D_p^{TOT})}{q(\mu_n N + \mu_p P)}\frac{\nabla^2(\Delta N)}{\Delta N} + \frac{\epsilon_0\epsilon_{\text{lattice}}k_b(\mu_n N - \mu_p P)}{q^2(\mu_n N + \mu_p P)}\frac{\nabla^2 T}{\Delta N}\right|. \tag{4.14}$$

The first term of the right hand-side of Eq. (4.14) corresponds to the term discussed by McKelvey and has been checked to always be small in Si. The second term is also small if, additionally,

$$\frac{\epsilon_0\epsilon_{\text{lattice}}k_b}{q^2}\frac{|\nabla^2 T|}{\Delta N} \ll 1, \tag{4.15}$$

Assuming an excess temperature decaying exponentially with a characteristic length L_d^{th} (see Sect. 4.2), the above expression becomes

$$\frac{L_{\text{Debye}}^2}{(L_d^{th})^2}\frac{|N - P|}{\Delta N}\frac{\Delta T}{T} \ll 1. \tag{4.16}$$

Owing to the thermal diffusion length being significantly longer than the Debye length in TP (μm vs nm), the condition of Eq. (4.16) is always verified ($<10^{-4}$ even in the worst case of highly doped substrates). In other words, the concentration disparity $(\Delta P - \Delta N)$ needed to create the internal field is much smaller than the concentrations ΔP and ΔN themselves, which ensures that charge balance is a good approximation even in the non-isothermal situation studied in this work.

Second, we assume that the BGN currents are only due to the generated free carriers. This is obviously valid in a homogeneously doped semiconductor sample. In this case, following Refs. [10, 11], $\vec{\nabla}\chi = \partial\chi/\partial(\Delta N)\vec{\nabla}(\Delta N)$ and $\vec{\nabla}(\chi + E_g) = \partial(\chi + E_g)/\partial(\Delta N)\vec{\nabla}(\Delta N)$, and the current Eq. (4.8) can be rewritten

$$\vec{J}_n = -q\mu_n N\vec{\nabla}\psi + qD_n^{TOT}\vec{\nabla}N + k_b\mu_n N\vec{\nabla}T \tag{4.17a}$$

$$\vec{J}_p = -q\mu_p P\vec{\nabla}\psi - qD_p^{TOT}\vec{\nabla}P - k_b\mu_p P\vec{\nabla}T, \tag{4.17b}$$

with the total diffusivities

$$D_n^{TOT} = D_n \underbrace{-\frac{N\mu_n}{q}\frac{\partial \chi}{\partial(\Delta N)}}_{D_n^{BGN}} \tag{4.18a}$$

$$D_p^{TOT} = D_p + \frac{P\mu_p}{q}\frac{\partial(\chi + E_g)}{\partial(\Delta N)} = D_p \underbrace{-\frac{P\mu_p}{q}\left|\frac{\partial(\chi + E_g)}{\partial(\Delta N)}\right|}_{D_p^{BGN}}. \tag{4.18b}$$

The two derivatives in Eqs. (4.18a, 4.18b) can be expressed using e.g. Schenk's BGN model [12]. BGN quasi-electric fields act therefore as counter-diffusive terms. The involved additional terms D_n^{BGN} and D_p^{BGN} are indeed always negative, hence reducing the total carrier diffusivities.

Third, we add together Eqs. (4.6b) and (4.6c) respectively multiplied by the hole conductivity $q\mu_p P$ and electron conductivity $q\mu_n N$, using the definition Eqs. (4.3), (4.4) and the current Eqs. (4.17a, 4.17b). This gives the crucial *generalized ambipolar diffusion equation*

$$\begin{aligned}\frac{\partial \Delta N}{\partial t} &= \mu_a \vec{\nabla}(\Delta N).\vec{\nabla}\psi + D_a \nabla^2(\Delta N) + G - \text{Rec} \\ &+ \frac{k_b}{q}\frac{\mu_n\mu_p(N+P)}{\mu_n N + \mu_p P}\vec{\nabla}(\Delta N).\vec{\nabla}T + \frac{k_b}{q}\frac{2\mu_n\mu_p NP}{\mu_n N + \mu_p P}\nabla^2 T,\end{aligned} \tag{4.19}$$

where

$$\mu_a = \frac{\mu_n\mu_p(N-P)}{\mu_n N + \mu_p P} \tag{4.20a}$$

$$D_a = \frac{\mu_p P D_n^{TOT} + \mu_n N D_p^{TOT}}{\mu_n N + \mu_p P}, \tag{4.20b}$$

are respectively the ambipolar mobility and ambipolar diffusivity. It has been assumed that $\vec{\nabla}(D_i\vec{\nabla}\Delta N) = D_i\nabla^2(\Delta N)$ ($i = n$ or p) since the variations in carrier diffusivities are much smoother than the variations in carrier concentrations (see e.g. Fig. 4.15 of Ref. [13]). Equation (4.19) contains all the carrier transport information needed for the understanding of a PMOR measurement, not only for silicon but also for other materials where electrothermal ambipolar motion of electrons and holes is allowed.

The fourth and final simplification consists in neglecting three of the terms in formula (4.19).

First, we neglect the first (drift) term of the right-hand side of Eq. (4.19). McKelvey shows that the drift effect related to the internal field is considerably smaller than diffusive effects in the isothermal case. In the non-isothermal case, we must check that similar conclusions can be reached about the internal field given by Eq. (4.12).

We must have

$$\left|\mu_a \vec{\nabla}(\Delta N).\left[\frac{(D_n^{TOT} - D_p^{TOT})}{(\mu_n N + \mu_p P)}\vec{\nabla}(\Delta N) + \frac{k_b(\mu_n N - \mu_p P)}{q(\mu_n N + \mu_p P)}\vec{\nabla}T\right]\right| \ll \left|D_a \nabla^2(\Delta N)\right|. \tag{4.21}$$

This is verified if we have simultaneously

$$|\Delta N| \ll \left|\frac{(\mu_n N + \mu_p P)}{(\mu_n - \mu_p)}\frac{(N+P)}{(P-N)}\right|, \tag{4.22}$$

and

$$|\Delta T| \ll T\frac{L_d^{th}}{L_d^{pl}}\left|\frac{(\mu_n N + \mu_p P)}{(\mu_n N - \mu_p P)}\frac{(N+P)}{(P-N)}\right|. \tag{4.23}$$

The derivation of Eqs. (4.22) and (4.23) assumes Boltzmann statistics and exponentially decaying carrier and temperature distributions with respective decay lengths L_d^{pl} and L_d^{th} (see e.g. Sect. 4.2). As checked by McKelvey (p. 330 of Ref. [9]), the first condition is always verified in Si. The second condition is valid, owing to the small magnitude of the temperature perturbations (less than one percent of T). We show in Sect. 4.2 that this is indeed always the case for a silicon substrate in the irradiance range of TP. For other highly absorptive materials like Ge, however, this term may be expected to be significant.

Second, we neglect the last two (Seebeck) terms of the right-hand side of Eq. (4.19), respectively proportional to the gradient $\vec{\nabla}T$ and laplacian $\nabla^2 T$ of the excess temperature. Let us assume that electrons and holes have equal mobility μ and equal diffusivity $D = k_b T\mu/q$. In this case, we have for the $\vec{\nabla}T$ term

$$\frac{k_b}{q}\frac{\mu_n\mu_p(N+P)}{\mu_n N + \mu_p P}\vec{\nabla}(\Delta N).\vec{\nabla}T = \frac{k_b}{q}\mu\vec{\nabla}(\Delta N).\vec{\nabla}T \approx D\vec{\nabla}(\Delta N).\frac{\vec{\nabla}T}{T}. \tag{4.24}$$

Once again, the small magnitude ($<1\,\%$) of the considered temperature perturbations renders this term negligible compared to the diffusive term.

Finally, for the $\nabla^2 T$ term, similar reasoning could be used so as to show that it is usually negligible. However, this conclusion can be reached in a more elegant way. Using the time-independent heat equation (see e.g. Sect. 4.1.2), one can show that

$$\nabla^2 T = -\frac{(h\nu_{\text{pump}} - E_g)}{k_{th}}G - \frac{E_g}{k_{th}}\text{Rec}. \tag{4.25}$$

Assuming equal mobility and diffusivity for both types of carriers, we can therefore deduce that

$$\frac{k_b}{q}\frac{2\mu_n\mu_p NP}{\mu_n N+\mu_p P}\nabla^2 T \approx -\frac{2}{1/P+1/N}\left(\underbrace{\frac{D(h\nu_{\text{pump}}-E_g)}{k_{th}T}}_{\approx 10^{-21}\ \text{cm}^3} G-\underbrace{\frac{DE_g}{k_{th}T}}_{\approx 10^{-21}\ \text{cm}^3}\text{Rec}\right). \tag{4.26}$$

In other words, this term reduces the carrier generation and enhances the carrier recombinations. In the case of silicon, however, it is smaller than 10 % of the carrier recombination and generation rates if the total electron or hole concentration is lower than roughly $10^{20}\,\text{cm}^{-3}$. It is therefore neglected in this work. This term in $\nabla^2 T$ should, however, be taken into account at very high doping or injection of in the case of a poor thermal conductor. These results are in agreement with the conclusions of Ref. [1], which confirms that the Seebeck effect is experimentally never observed and therefore negligible compared to the diffusive term and the recombination/generation rates.

4.1.2 Heat Equation

As already mentioned, our main assumption concerning the energy transport is the equality of the carrier and lattice temperatures. As a consequence, the problem is summed up by a heat equation with five generation terms (Eq. (4.9)).

The first heat generation term is the direct heating (hot-carrier thermalization). This contribution occurs before any carrier transport and reads

$$G_{th}^{\text{direct}} = (h\nu_{\text{pump}} - E_g)G, \tag{4.27}$$

where empty bands are assumed. This term is not included in Kells's model since Kells considers that the carriers are constantly in thermal equilibrium with the lattice. This is obviously not possible in the case of optical generation since the carriers are initially generated with $(h\nu_{\text{pump}} - E_g)$ energy in excess to the bandgap. This excess energy is directly (after a few picoseconds [4]) released to the lattice, explaining the presence of this extra term. This term is usually not included in commercial numerical simulation software packages, which makes their use impossible when studying PMOR. This is the main reason for which a dedicated numerical simulation code has been developed [14] (FSEM). Note that, in theory, the direct heating also includes a contribution due to free-carrier absorption. This effect is, however, considerably smaller than the hot-carrier thermalization effect and is therefore neglected [13].

Further, the second generation term is the recombination heat

$$G_{th}^{\text{recombination}} = \left[E_f^n - E_f^p + qT(P_p - P_n)\right]\text{Rec}, \tag{4.28}$$

where the electron and hole thermoelectric powers assuming Boltzmann statistics are respectively [3] $P_n = -k_b/q\,[5/2 - ln(N/N_c)]$ and $P_p = k_b/q\,[5/2 - ln(P/N_v)]$ with N_c and N_v the effective densities of state of respectively the conduction and valence bands. Injecting these formulas into (4.28) gives

$$G_{th}^{\text{recombination}} = (E_g + 5k_bT)\text{Rec}. \tag{4.29}$$

Using Boltzmann statistics in the definition of the thermoelectric powers has implicitly modified the impact of band-filling by the excess carriers. As mentioned for the direct heating, we consider empty bands when it comes to heat generation. As the $5k_bT$ term of formula (4.29) is related to band-filling, it has to be ignored. Further, we observe below that this term needs to be neglected for the sake of energy conservation. This is consistent with the usual modeling [1, 15].

The next three generation terms are usually not taken into account and are all proportional to a power of the currents. It is easy to show that they all are negligible at room temperature. Using the one-dimensional linear model which will be developed in Sect. 4.2.1, assuming low-frequency diffusive currents (i.e. $J_n = -J_p = -q\sqrt{D_a/\tau}\Delta N$ where τ is the carrier recombination lifetime) and equal hole and electron mobilities, one finds that

$$|G_{th}^{\text{Joule}}| = \left|\frac{|\vec{J}_n|^2}{q\mu_n N} + \frac{|\vec{J}_p|^2}{q\mu_p P}\right| \leq k_bT\left(\frac{\Delta N}{N} + \frac{\Delta P}{P}\right)\text{Rec} \leq (2k_bT)\text{Rec} \tag{4.30a}$$

$$|G_{th}^{\text{Peltier}}| = \left|-T\left(\vec{J}_n.\frac{\partial P_n}{\partial N}\vec{\nabla}N + \vec{J}_p.\frac{\partial P_p}{\partial P}\vec{\nabla}P\right)\right| \leq k_bT\left(\frac{\Delta N}{N} + \frac{\Delta P}{P}\right)\text{Rec} \leq (2k_bT)\text{Rec} \tag{4.30b}$$

$$|G_{th}^{\text{Thomson}}| = \left|-T\left(\vec{J}_n.\frac{\partial P_n}{\partial T}\vec{\nabla}T + \vec{J}_p.\frac{\partial P_p}{\partial T}\vec{\nabla}T\right)\right| \leq \left[k_b\Delta T\left|ln\left(\frac{NP}{N_cN_v}\right)\sqrt{\frac{D_a}{2D_{th}}\omega\tau}\right|\right]\text{Rec}, \tag{4.30c}$$

where ω is the pump angular modulation frequency. At room temperature, these three heat generation terms account therefore for only a few percent of the recombination heat and can therefore be neglected.

Note that the total energy of the system is conserved. The sum of the direct heating and recombination contributions integrated over the sample volume is equal to the incoming optical (non-reflected) energy. Kells shows that conservation of energy is also ensured in a more general case [3].

4.1.3 Steady-Periodic Model Equations

In summary, based on the considerations of Sects. 4.1.1 and 4.1.2, the simplified coupled system of equations to be solved is the following

$$\frac{\partial \Delta N}{\partial t} = D_a\nabla^2(\Delta N) + G^{\text{pump}} + G^{\text{probe}} - \text{Rec} \tag{4.31a}$$

$$\rho c_p\frac{\partial T}{\partial t} = k_{th}\nabla^2 T + (h\nu_{\text{pump}} - E_g)G^{\text{pump}} + (h\nu_{\text{probe}} - E_g)G^{\text{probe}} + E_g\text{Rec}, \tag{4.31b}$$

where G^{pump} and G^{probe} are the carrier generation rates respectively due to the pump and the probe lasers.

Since we are dealing with *modulated* distributions of the carrier concentration and the temperature (Eqs. (4.3)–(4.5)], it is convenient to write directly the equations for the different harmonics of these distributions. Plugging Eq. (4.3) into Eq. (4.31a) and relying on the orthogonality of the exponential functions, we obtain the following steady(-periodic) equations to be solved for the different Fourier coefficients of the carrier concentration and temperature

$$ji\omega\Delta N_j = [D_a\nabla^2(\Delta N)]_j + G_j^{\text{pump}} + G_j^{\text{probe}} - \text{Rec}_j \tag{4.32a}$$

$$\begin{aligned} ji\omega\Delta T_j &= [D_{th}\nabla^2(\Delta T)]_j \\ &\quad + 1/(\rho c_p)\left\{[(h\nu_{\text{pump}} - E_g)G^{\text{pump}}]_j + [(h\nu_{\text{probe}} - E_g)G^{\text{probe}}]_j \right. \\ &\quad \left. +[E_g\text{Rec}]_j\right\} \\ &\quad \text{for } j = 0, 1, 2, 3, \ldots, \end{aligned} \tag{4.32b}$$

where $D_{th} = k_{th}/(\rho c_p)$ is the sample thermal diffusivity. In the ambipolar diffusion Eq. (4.32a), we have assumed the following Fourier expansions respectively of the carrier generation rate $G^{\text{laser}}(t)$ (laser=probe or pump), of the carrier diffusion rate $[D_a\nabla^2(\Delta N)](t)$ and of the carrier recombination rate $\text{Rec}(t)$

$$\begin{aligned} G^{\text{laser}}(t) &= G_0^{\text{laser}} \\ &+ \sum_{j=1}^{+\infty}[G_j^{\text{laser}}\exp(ij\omega t) + (G_j^{\text{laser}})^*\exp(-ij\omega t)] \end{aligned} \tag{4.33a}$$

$$\begin{aligned} [D_a\nabla^2(\Delta N)](t) &= [D_a\nabla^2(\Delta N)]_0 \\ &+ \sum_{j=1}^{+\infty}\left\{[D_a\nabla^2(\Delta N)]_j\exp(ij\omega t) + [D_a\nabla^2(\Delta N)]_j^*\exp(-ij\omega t)\right\} \end{aligned} \tag{4.33b}$$

$$\begin{aligned} \text{Rec}(t) &= \text{Rec}_0 \\ &+ \sum_{j=1}^{+\infty}[\text{Rec}_j\exp(ij\omega t) + \text{Rec}_j^*\exp(-ij\omega t)]. \end{aligned} \tag{4.33c}$$

Similarly, in the heat Eq. (4.32b), we have assumed the following Fourier expansions respectively of the direct heating $[(h\nu_{\text{laser}} - E_g)G^{\text{laser}}](t)$ (laser=probe or pump), of the recombination heat $[E_g\text{Rec}](t)$ and of the heat diffusion rate $[D_{th}\nabla^2 (\Delta T)](t)$

$$[(h\nu_{\text{laser}} - E_g)G^{\text{laser}}](t) = [(h\nu_{\text{laser}} - E_g)G^{\text{laser}}]_0$$

$$+\sum_{j=1}^{+\infty}\left\{[(h\nu_{\text{laser}}-E_g)G^{\text{laser}}]_j\exp(ij\omega t)+[(h\nu_{\text{laser}}-E_g)G^{\text{laser}}]_j^*\exp(-ij\omega t)\right\}$$

(4.34a)

$$[E_g\text{Rec}](t)=[E_g\text{Rec}]_0$$

$$+\sum_{j=1}^{+\infty}\left\{[E_g\text{Rec}]_j\exp(ij\omega t)+[E_g\text{Rec}]_j^*\exp(-ij\omega t)\right\} \tag{4.34b}$$

$$[D_{th}\nabla^2(\Delta T)](t)=[D_{th}\nabla^2(\Delta T)]_0$$

$$+\sum_{j=1}^{+\infty}\left\{[D_{th}\nabla^2(\Delta T)]_j\exp(ij\omega t)+[D_{th}\nabla^2(\Delta T)]_j^*\exp(-ij\omega t)\right\}. \tag{4.34c}$$

Equations (4.32a, 4.32b) together with Eqs. (4.33a–4.33c) and (4.34a–4.34c) are the model equations for the calculations of the Fourier coefficients of the carrier concentration and temperature. As such, we have an infinite system of coupled nonlinear equations to be solved. However, as we show below, this system can be greatly simplified in the specific case of TP on silicon.

The key to simplifying the problem lies in the evaluation of the efficiency of the nonlinearity of our model Eq. (4.32a, 4.32b) to *couple* and to *generate harmonics*. The efficiencies of all the nonlinear effects should therefore be examined. We start with the nonlinear carrier generation rate in Sect. 4.1.3.1 and proceed with the nonlinear carrier diffusion rate in Sect. 4.1.3.2. The nonlinear carrier recombination rate is then investigated in Sect. 4.1.3.3. Though the heat Eq. (4.32b) are expected to be linear, owing to the small magnitudes of the temperature perturbations, we briefly discuss the heat generation rate in Sect. 4.1.3.4 and the heat diffusion rate in Sect. 4.1.3.5.

As it turns out, we shall show that the harmonic-generation efficiency of these nonlinear effects is negligible in our study. Only four equations among (4.32a, 4.32b) will therefore remain (ΔN_0, ΔN_1, ΔT_0 and ΔT_1). These equations will, however, be coupled by all the nonlinear effects and predominantly by the nonlinear diffusion and recombination rates.

To prove these statements, we propose to take a case study. In theory, both the cases of low and high carrier injections should be discussed. However, in case of low injection, all coefficients of the system become independent from the excess carrier concentration. The system retrieving its linearity, the low-injection case is trivial and does not need any further discussion. We therefore focus on the case of high carrier injection. More particularly, we consider the typical example of a lowly doped p-type silicon sample ($P_{\text{doping}}=10^{15}\,\text{cm}^{-3}$) illuminated by a supra-bandgap laser with the following modulated irradiance

$$\mathcal{P}_{\text{laser}}(t)=\frac{\mathcal{P}^0_{\text{laser}}}{2}+\frac{\mathcal{P}^0_{\text{laser}}}{4}\exp(i\omega t)+\frac{\mathcal{P}^0_{\text{laser}}}{4}\exp(-i\omega t). \tag{4.35}$$

We suppose that only the constant and fundamental modes of the excess carrier distribution are excited, i.e.

$$\begin{aligned}\Delta N(x, y, z, t) &= \Delta N_{\text{ref}}(x, y, z)[1 + \cos(\omega t)] \\ &= \Delta N_{\text{ref}}(x, y, z)[1 + 0.5 \exp(i\omega t) + 0.5 \exp(-i\omega t)], \end{aligned} \tag{4.36}$$

with $\Delta N_{\text{ref}}(0, 0, 0) = 10^{18}\,\text{cm}^{-3}$. The considered variation with time in carrier concentration is depicted in Fig. 4.1 both in the time domain (Fig. 4.1a) and in the frequency domain[2] (Fig. 4.1b).

Note, finally, that this study is focused on the behavior at position $(x, y, z) = (0, 0, 0)$. The spatial dependence is therefore not explicitly written in the rest of this study.

4.1.3.1 Nonlinear Carrier Generation Rate

The optical carrier generation rate[3] G of a supra-bandgap laser follows Beer's law [17], i.e.

$$G(x, y, z, t) = \alpha_{\text{BTB}}^{\text{laser}}(1 - R_0)\mathcal{P}_{\text{laser}}(t) \exp\left(-\frac{x^2 + y^2}{\mathcal{R}_{\text{laser}}^2}\right) \exp(-\alpha_{\text{BTB}}^{\text{laser}} z)/(h\nu_{\text{laser}}), \tag{4.37}$$

where $\alpha_{\text{BTB}}^{\text{laser}}$ is the band-to-band absorption coefficient of silicon at the optical frequency ν_{laser} of the laser. $\mathcal{P}_{\text{laser}}(t)$ is the laser irradiance, $\mathcal{R}_{\text{laser}}$ is the radius of the gaussian laser beam. Rigorously, R_0 should actually be replaced by R in Eq. (4.37), i.e. the perturbed reflectance (Chap. 2), but the small amplitude of the considered reflectance perturbations can here be neglected.

The band-to-band absorption coefficient $\alpha_{\text{BTB}}^{\text{laser}}$ is the most complex parameter of Eq. (4.37). It has already been discussed in Sect. 3.2.2 where we show that it not only varies with the laser wavelength but also with the sample bandgap and therefore carrier concentration. In other words, owing to the presence of $\alpha_{\text{BTB}}^{\text{laser}}$ in Eq. (4.37), the optical carrier generation rate is nonlinear. To model this effect, we make use of Schenk's BGN model [12] and Smith's band-to-band absorption model [18]. The reason for using Smith's band-to-band absorption model is that, contrary to most models, the bandgap energy appears directly in this model. It is therefore possible to account for the variations in absorption coefficient with carrier-induced BGN. Note, however, that the combination of these models tends to underestimate the band-to-band absorption coefficient compared to experimental values [19, 20]. This is illustrated in Fig. 4.2, where it is observed that the values of the bandgap needed

---

[2] Since we only deal with real distributions in this work, the negative frequencies are never discussed or represented. They can of course be deduced by symmetry with respect to $j = 0$

[3] The only carrier generation rate considered in our investigations is the optical generation rate. We neglect the thermal generation of carriers in spite of the heating occuring under the laser beams. This can be justified by the low excess temperatures considered in this study (see Sect. 4.2 and Ref. [16]).

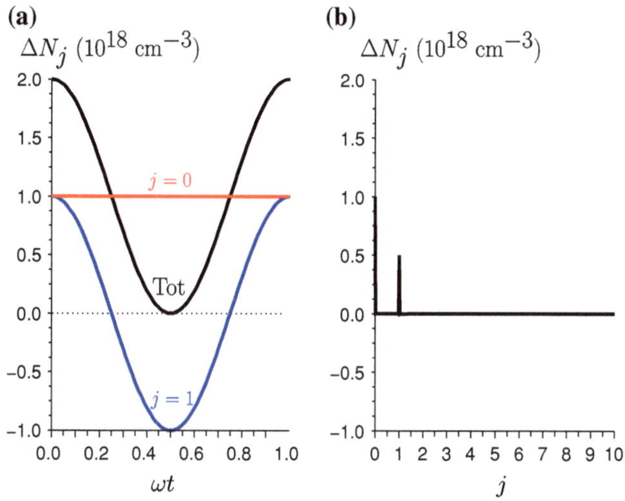

Fig. 4.1 **a** Time-domain representation of the considered example carrier concentration (*black curve*) and of the various harmonics composing it (*colored curves*). **b** Frequency-domain representation of the considered example carrier concentration. The Fourier coefficients corresponding to negative frequencies can be deduced by symmetry with respect to $j = 0$

to fit the experimental α_{BTB} with Smith's model are always smaller than expected from Schenk's model. We will come back to this in Chap. 6.

Based on Schenk's and Smith's models, we would like to evaluate the *coupling* and *harmonic-generation* efficiencies of the nonlinear carrier generation rate. In order to do so, we first need to look at the time dependence of the bandgap and of the subsequent band-to-band absorption coefficient, i.e.

$$E_g(t) = E_{g0} + \sum_{j=1}^{+\infty} [E_{gj} \exp(ij\omega t) + (E_{gj})^* \exp(-ij\omega t)] \tag{4.38a}$$

$$\alpha_{\mathrm{BTB}}^{\mathrm{laser}}(t) = \alpha_{\mathrm{BTB}_0}^{\mathrm{laser}} + \sum_{j=1}^{+\infty} [\alpha_{\mathrm{BTB}_j}^{\mathrm{laser}} \exp(ij\omega t) + (\alpha_{\mathrm{BTB}_j}^{\mathrm{laser}})^* \exp(-ij\omega t)]. \tag{4.38b}$$

To quantify each Fourier coefficient of Eqs. (4.38a) and (4.38b), we plug the excess carrier concentration of Eq. (4.36) into Schenk's BGN model and Smith's absorption model. We obtain the Fourier series of E_g and $\alpha_{\mathrm{BTB}}^{\mathrm{laser}}$ shown in Fig. 4.3a and b respectively. More specifically, these Figures show the values of the various Fourier components normalized by their constant mode. From these Figures, it appears that the constant mode of both the bandgap and the absorption coefficient is two to three

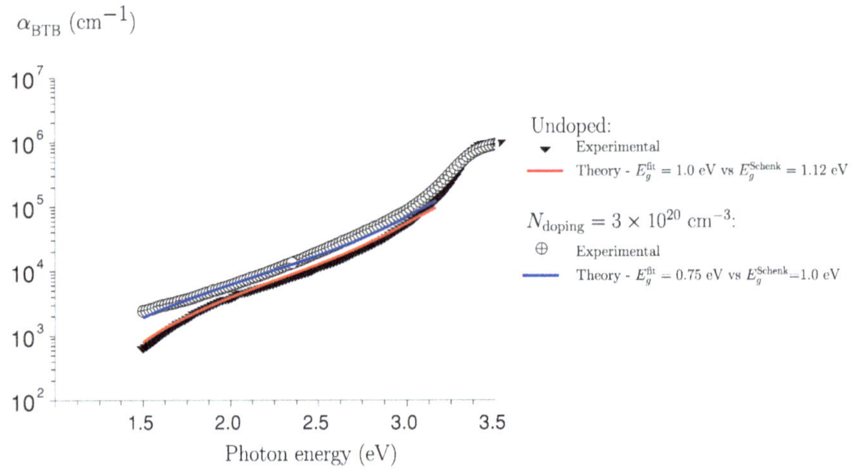

Fig. 4.2 Variation in band-to-band absorption coefficient of Si with photon energy for undoped and highly doped Si (3×10^{20} cm^{-3}). The *black triangles* and the *crossed circles* show experimental absorption coefficients measured respectively on undoped and highly doped Si [19, 20]. The *red* and *blue curves* are theoretical lines given by Smith's band-to-band absorption model respectively for $E_g = 1.0$ eV and $E_g = 0.75$ eV. The values of the bandgap E_g^{fit} used to fit the experimental data with Smith's model are lower than Schenk's bandgap values E_g^{Schenk}. The combination of Smith's and Schenk's models therefore underestimates α_{BTB}

orders of magnitude larger than all the other Fourier components, i.e. E_g and $\alpha_{\mathrm{BTB}}^{\mathrm{laser}}$ hardly vary with time at all.

It is therefore a very good approximation to consider only the constant modes of the bandgap and absorption coefficient, which leads to the following Fourier expansion of the carrier generation rate

$$\begin{aligned} G_0 &= (1 - R_0)/(h\nu_{\mathrm{laser}}) \left[\alpha_{\mathrm{BTB}_0}^{\mathrm{laser}} \mathcal{P}_{\mathrm{laser}}^0/2 + \Re\left(\alpha_{\mathrm{BTB}_1}^{\mathrm{laser}} \mathcal{P}_{\mathrm{laser}}^0/2\right)\right] \\ &\approx (1 - R_0)/(h\nu_{\mathrm{laser}})(\alpha_{\mathrm{BTB}_0}^{\mathrm{laser}} \mathcal{P}_{\mathrm{laser}}^0/2) \qquad (4.39a) \\ G_1 &= (1 - R_0)/(h\nu_{\mathrm{laser}})(\alpha_{\mathrm{BTB}_0}^{\mathrm{laser}} \mathcal{P}_{\mathrm{laser}}^0/4 + \alpha_{\mathrm{BTB}_1}^{\mathrm{laser}} \mathcal{P}_{\mathrm{laser}}^0/2 + \alpha_{\mathrm{BTB}_2}^{\mathrm{laser}} \mathcal{P}_{\mathrm{laser}}^0/4) \\ &\approx (1 - R_0)/(h\nu_{\mathrm{laser}})(\alpha_{\mathrm{BTB}_0}^{\mathrm{laser}} \mathcal{P}_{\mathrm{laser}}^0/4) \qquad (4.39b) \\ G_j &= (1 - R_0)/(h\nu_{\mathrm{laser}})(\alpha_{\mathrm{BTB}_{j-1}}^{\mathrm{laser}} \mathcal{P}_{\mathrm{laser}}^0/4 + \alpha_{\mathrm{BTB}_j}^{\mathrm{laser}} \mathcal{P}_{\mathrm{laser}}^0/2 + \alpha_{\mathrm{BTB}_{j+1}}^{\mathrm{laser}} \mathcal{P}_{\mathrm{laser}}^0/4) \\ &\approx 0 \text{ for } j = 2, 3, \ldots. \qquad (4.39c) \end{aligned}$$

These Fourier coefficients, normalized by their constant mode G_0, are shown in Fig. 4.4, where it is obvious that the Fourier expansion can be limited to its constant and fundamental modes. In conclusion, the nonlinear carrier generation rate does not generate any harmonic.

Besides, the nonlinear carrier generation rate induces very limited coupling since only the knowledge of $\alpha_{\mathrm{BTB}_0}^{\mathrm{laser}}$ is necessary for the calculation of all components of

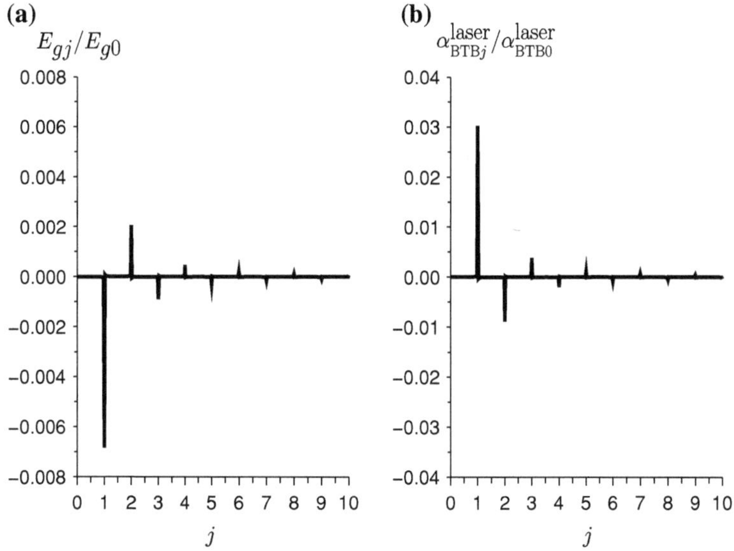

Fig. 4.3 **a** Normalized Fourier coefficients of the bangap when a harmonic carrier concentration such as given by Eq. (4.36) perturbs the bandgap of a sample with $10^{15}\,\text{cm}^{-3}$ p-type doping. The normalization of the coefficients with the constant mode E_{g0} clearly highlights the dominance of the latter. These results were obtained using Schenk's BGN model [12]. **b** Subsequent normalized Fourier coefficients of the band-to-band absorption coefficient α_{BTB}. The normalization of the coefficients with the constant mode α_{BTB_0} also shows that the latter dominates. These results were obtained using Smith's band-to band absorption model [18]

the carrier generation rate. It, however, has to be kept in mind that this calculation requires the knowledge of all Fourier components of the carrier distribution. Though limited, this coupling still needs to be taken into account.

4.1.3.2 Nonlinear Carrier Diffusion Rate

The carrier diffusion rate $D_a \nabla^2 \Delta N$ presents a nonlinear behavior due to the dependence of the ambipolar diffusivity D_a on carrier concentration. The expression of the ambipolar diffusivity can be derived from Eqs. (4.20b) and (4.18a, 4.18b) complemented with the Einstein relationships linking carrier mobilities and diffusivities, i.e.

$$D_n = \frac{k_b T}{q} \frac{F_{1/2}\left(\frac{E_f^n - E_c}{k_b T}\right)}{F_{-1/2}\left(\frac{E_f^n - E_c}{k_b T}\right)} \mu_n \tag{4.40a}$$

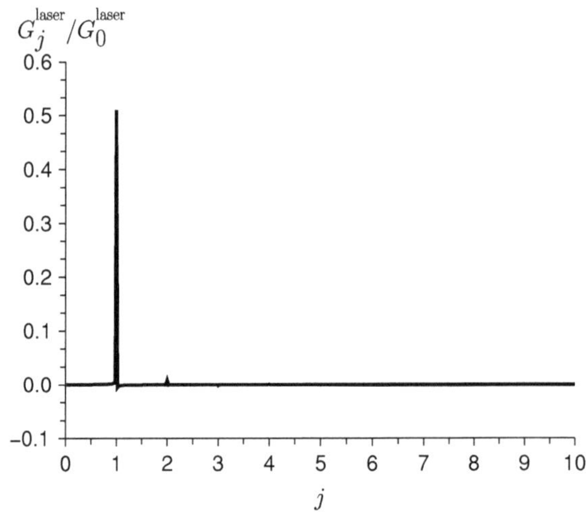

Fig. 4.4 Normalized Fourier coefficients of the optical carrier generation when a laser with irradiance such as given by Eq. (4.35) shines on a sample with a harmonic carrier concentration such as given by Eq. (4.36). In spite of its nonlinearity, the optical carrier generation rate only presents a constant mode and a fundamental mode

$$D_p = \frac{k_b T}{q} \frac{F_{1/2}\left(\frac{E_v - E_f^p}{k_b T}\right)}{F_{-1/2}\left(\frac{E_v - E_f^p}{k_b T}\right)} \mu_p. \tag{4.40b}$$

where E_c and E_v are respectively the conduction band and the valence band edges, $F_{1/2}$ and $F_{-1/2}$ are the Fermi integrals as defined in Ref. [5, 6]. To acount for the dependence of the electron and hole mobilities upon the carrier concentration, we make use of Klaassen's model [21]. This model accurately accounts for the variations in mobility both with doping and carrier injection and is therefore well suited for this study.

In summary, the dependence of the ambipolar diffusivity upon carrier concentration is quite complex since the carrier concentration is involved at different levels of the modeling of this transport parameter. It can be checked in Fig. 4.5 that for lowly doped p- or n-type silicon substrates, our model of the ambipolar diffusivity nicely follows the experimental behavior presented in Ref. [22]. Note, however, that the BGN-dependence of the ambipolar diffusivity, embodied by D_n^{BGN} and D_p^{BGN}, has not been accounted for in the theoretical curves shown in Fig. 4.5. We will come back to this remark in Chap. 6.

Based on this model, we would like to investigate the coupling and harmonic-generation efficiency of the nonlinear carrier diffusion rate. Let us first write the Fourier expansion of the time-dependent electron mobility, hole mobility and ambipolar diffusivity, i.e.

$$\mu_n(t) = \mu_{n0} + \sum_{j=1}^{+\infty} [\mu_{nj} \exp(ij\omega t) + (\mu_{nj})^* \exp(-ij\omega t)] \tag{4.41a}$$

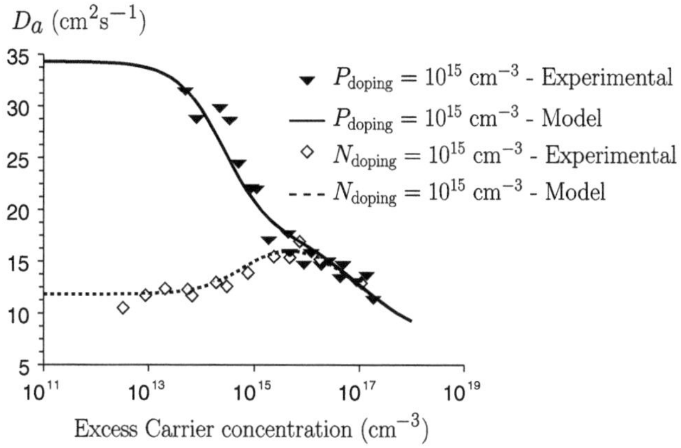

Fig. 4.5 Variation in ambipolar diffusivity with carrier injection in lowly doped n- and p-type Si. The theoretical lines are in excellent agreement with the experimental symbols of Ref. [22]. The theoretical lines assume $D_n^{BGN} = D_p^{BGN} = 0$

$$\mu_p(t) = \mu_{p0} + \sum_{j=1}^{+\infty} [\mu_{pj} \exp(ij\omega t) + (\mu_{pj})^* \exp(-ij\omega t)] \tag{4.41b}$$

$$D_a(t) = D_{a0} + \sum_{j=1}^{+\infty} [D_{aj} \exp(ij\omega t) + (D_{aj})^* \exp(-ij\omega t)]. \tag{4.41c}$$

The Fourier coefficients of these three expansions, normalized by their respective constant modes, are shown in Fig. 4.6 for the specific case of our example carrier concentration given in Eq. (4.36). It is clearly observed that, for all three transport parameters, though the constant mode dominates, the harmonics are not negligible. Most importantly, the fundamental mode of D_a is only about five times smaller than its constant mode.

Taking these results into account, the Fourier coefficients of the diffusion rate of Eq. (4.33b) can be written in the case of our example carrier concentration

$$[D_a \nabla^2(\Delta N)]_0 = D_{a0} \nabla^2(\Delta N_0) + 2\Re[D_{a1}^* \nabla^2(\Delta N_1)] \tag{4.42a}$$

$$[D_a \nabla^2(\Delta N)]_1 = D_{a0} \nabla^2(\Delta N_1) + D_{a1} \nabla^2(\Delta N_0) + D_{a2} \nabla^2(\Delta N_1^*) \tag{4.42b}$$

$$[D_a \nabla^2(\Delta N)]_j = D_{a(j-1)} \nabla^2(\Delta N_1) + D_{aj} \nabla^2(\Delta N_0) + D_{a(j+1)} \nabla^2(\Delta N_1^*) \quad \text{for } j = 2, 3, \ldots. \tag{4.42c}$$

The first term of each Eq. (4.42a–4.42c) dominates but the other terms cannot be neglected. This very important result shows that the nonlinear carrier diffusion rate acts as an efficient frequency mixer (equation coupling). As can be observed in Fig. 4.7 showing its normalized Fourier coefficients, however, the nonlinear carrier diffusion rate does not generate harmonics.

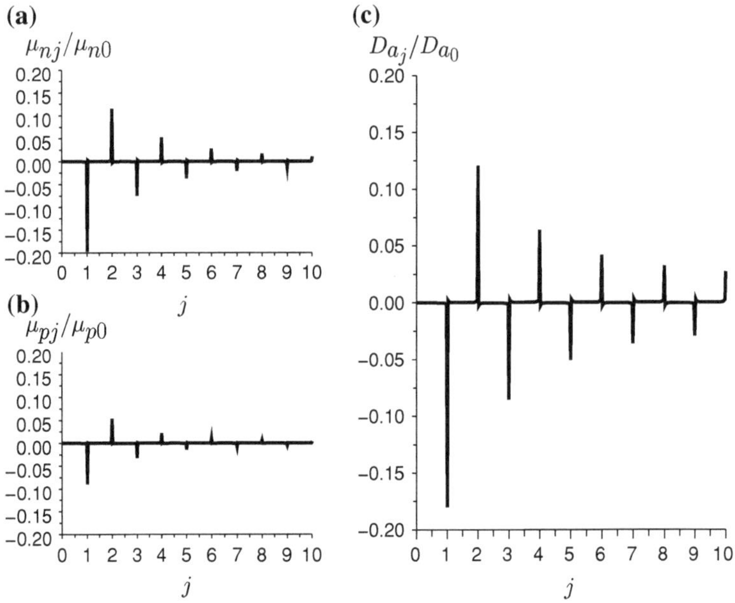

Fig. 4.6 **a** Normalized Fourier coefficients of the electron mobility μ_n when a harmonic carrier concentration such as given by Eq. (4.36) perturbs a sample with 10^{15} cm^{-3} p-type doping. **b** Normalized Fourier coefficients of the hole mobility μ_p in the same conditions. **c** Normalized Fourier coefficients of the subsequent ambipolar diffusivity D_a. The normalization of the coefficients with the constant mode D_{a0} highlights that, though D_{a0} dominates, the other Fourier coefficients of D_a are non negligible

4.1.3.3 Nonlinear Carrier Recombination Rate

In a semiconductor with an indirect bandgap like silicon, carriers can recombine via two independent phenomena. First, carriers can recombine via a Shockley-Read-Hall (SRH) mechanism [5], i.e. via a defect level located in the bandgap. Second, an electron-hole pair can recombine via Auger recombination, i.e. by scattering on a third free carrier. Notice that radiative recombinations, a third recombination mechanism, also exists but is very unlikely at room temperature [23] and is therefore not discussed here. In other words, the total recombination rate reads

$$\mathrm{Rec} = \mathrm{Rec}^{\mathrm{SRH}} + \mathrm{Rec}^{\mathrm{Auger}}, \tag{4.43}$$

where $\mathrm{Rec}^{\mathrm{SRH}}$ is the SRH recombination rate and $\mathrm{Rec}^{\mathrm{Auger}}$ is the Auger recombination rate.

The SRH recombination mechanism involves a free electron and a free hole. It is therefore easy to understand that the SRH recombination rate $\mathrm{Rec}_{\mathrm{SRH}}$ is proportional to the product of their concentrations. The SRH recombination can be shown to be [5, 13]

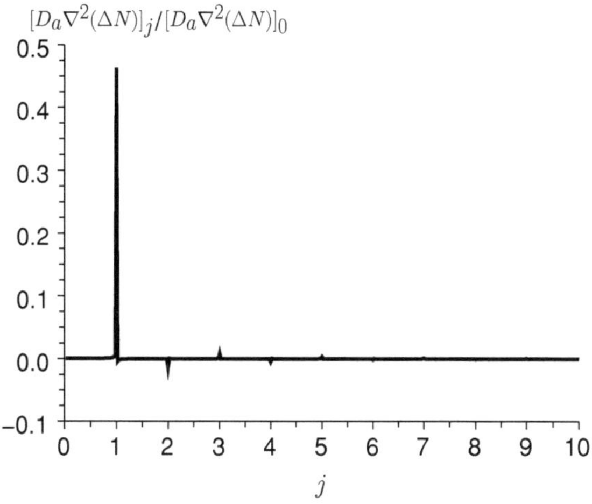

Fig. 4.7 Normalized Fourier coefficients of the carrier diffusion rate when a harmonic carrier concentration such as given by Eq. (4.36) perturbs a sample with $10^{15}\,\mathrm{cm}^{-3}$ p-type doping. In spite of its nonlinearity and the numerous harmonics of D_a, the carrier diffusion rate only presents significant constant and fundamental modes

$$\mathrm{Rec}^{\mathrm{SRH}} = \frac{PN - n_i^2\gamma_n\gamma_p}{\tau_p(N + n_i\gamma_n) + \tau_n(P + n_i\gamma_p)} \tag{4.44}$$

where n_i is the effective carrier density of silicon [3], γ_n (resp. γ_p) is the Fermi-Dirac factor for electrons (resp. for holes)

$$\gamma_n = \frac{F_{1/2}\left\{\left[E_f^n - E_c\right]/(k_bT)\right\}}{\exp\left\{\left[E_f^n - E_c\right]/(k_bT)\right\}} \tag{4.45a}$$

$$\gamma_p = \frac{F_{1/2}\left\{\left[E_v - E_f^p\right]/(k_bT)\right\}}{\exp\left\{\left[E_v - E_f^p\right]/(k_bT)\right\}}, \tag{4.45b}$$

τ_n (resp. τ_p) is the concentration-dependent SRH lifetime of electrons (resp. holes), i.e.

$$\tau_n = \frac{\tau_n^0}{1 + (N_a^- + N_d^+)/N_{\mathrm{SRH}}} \tag{4.46a}$$

$$\tau_p = \frac{\tau_p^0}{1 + (N_a^- + N_d^+)/N_{\mathrm{SRH}}}, \tag{4.46b}$$

where τ_n^0 (resp. τ_p^0) are the low-concentration SRH lifetime of electrons (resp. holes) and N_{SRH} is the so-called SRH doping concentration [5].

The Fourier expansion of the SRH recombination rate is readily derived assuming our example carrier concentration of Eq. (4.36)

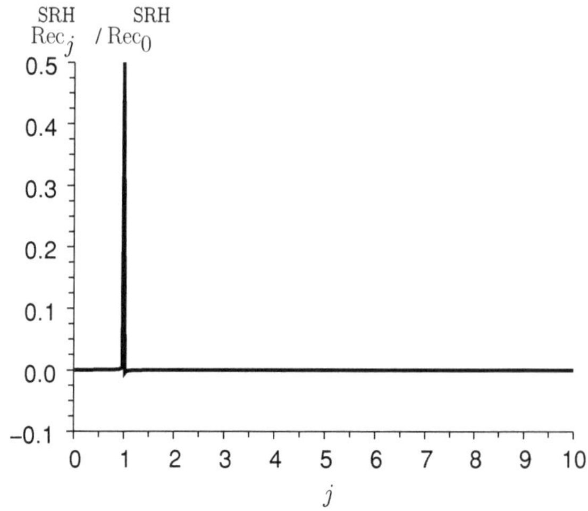

Fig. 4.8 Normalized Fourier coefficients of the SRH recombination rate when a harmonic carrier concentration such as given by Eq. (4.36) perturbs a sample with 10^{15} cm^{-3} p-type doping. The SRH recombination rate proves to be linear at high injection, thus generating no harmonics and inducing negligible coupling between the model equations

$$\mathrm{Rec}_0^{\mathrm{SRH}} \approx \frac{\Delta N_{\mathrm{ref}}}{(\tau_n + \tau_p)} \tag{4.47a}$$

$$\mathrm{Rec}_1^{\mathrm{SRH}} \approx \frac{\Delta N_{\mathrm{ref}}}{2(\tau_n + \tau_p)} \tag{4.47b}$$

$$\mathrm{Rec}_j^{\mathrm{SRH}} \approx 0 \text{ for } j = 2, 3, \ldots \tag{4.47c}$$

after neglecting P_{doping}, $n_i \gamma_n$ and $n_i \gamma$ before the excess carrier concentration[4]. In other words, the SRH recombination rate at high injection is linear and, therefore, generates no harmonics and no coupling. The Fourier expansion of the SRH recombinations is shown of Fig. 4.8.

The Auger recombination is a phenomenon which involves three charge carriers. It is therefore easy to understand that the Auger recombination rate involves third powers of the carrier concentrations. The Auger recombination rate indeed reads [5, 13]

$$\mathrm{Rec}^{\mathrm{Auger}} = C_n N (PN - n_i^2 \gamma_n \gamma_p) + C_p P (PN - n_i^2 \gamma_n \gamma_p), \tag{4.48}$$

where C_n and C_p are two constants [5].

Equation (4.48) shows that Auger recombinations are a strongly nonlinear phenomenon. It can therefore be expected that the Auger recombination rate generates harmonics. Similar to SRH recombinations, we can directly write the analytical expression of all the Fourier coefficients of the Auger recombination rate. In the case of our example carrier concentration of Eq. (4.36), these Fourier coefficients read[5]

4 See Appendix A.7 for a more general Fourier expansion of the SRH recombination rate

5 See Appendix A.7 for a more general Fourier decomposition of the Auger recombination rate

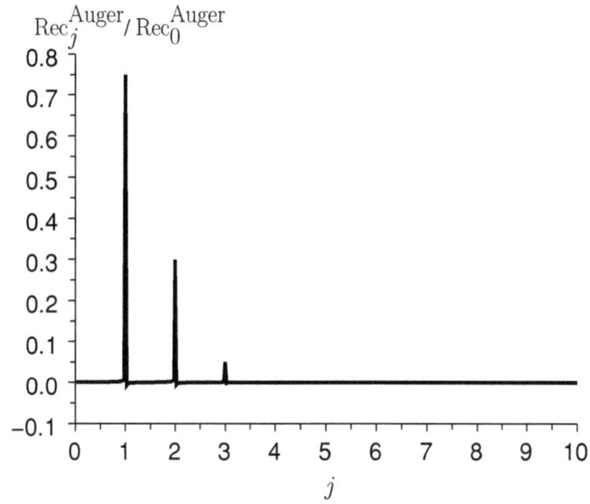

Fig. 4.9 Normalized Fourier coefficients of the Auger recombination rate when a harmonic carrier concentration such as given by Eq. (4.36) perturbs a sample with $10^{15}\,\mathrm{cm}^{-3}$ p-type doping. The nonlinearity of the Auger recombination rate at high injection generates second and third harmonics and coupling between the model equations

$$\mathrm{Rec}_0^{\mathrm{Auger}} = \frac{5}{2}(C_n + C_p)\Delta N_{\mathrm{ref}}^3 \tag{4.49a}$$

$$\mathrm{Rec}_1^{\mathrm{Auger}} = \frac{15}{8}(C_n + C_p)\Delta N_{\mathrm{ref}}^3 \tag{4.49b}$$

$$\mathrm{Rec}_2^{\mathrm{Auger}} = \frac{3}{4}(C_n + C_p)\Delta N_{\mathrm{ref}}^3 \tag{4.49c}$$

$$\mathrm{Rec}_3^{\mathrm{Auger}} = \frac{1}{8}(C_n + C_p)\Delta N_{\mathrm{ref}}^3 \tag{4.49d}$$

$$\mathrm{Rec}_j^{\mathrm{Auger}} = 0 \text{ for } j = 4, 5, \ldots \tag{4.49e}$$

The normalized Fourier coefficients are shown in Fig. 4.9. As could have been expected from the third-order polynomial form of Eq. (4.48), Auger recombinations generate second and third harmonics.

In summary, while SRH recombinations turn out to be linear at high injection, Auger recombinations are strongly nonlinear, hence inducing coupling and the creation of harmonics. Though this statement is correct in the most general case, we here need to remind one specificity of TP. The radii $\mathcal{R}_{\mathrm{laser}}$ of the probe and pump lasers of TP are very small (0.5 μm), which strongly favors diffusion over recombinations.[6] Consequently, the carrier transport is diffusion-limited up to very high doping concentrations in Si. In other words, though harmonics are indeed generated by Auger recombinations, their magnitude and the coupling they involve are negligible. This is in agreement with Fig. A.13 of Appendix A.7.1.

[6] As a rule of thumb, the relative importance of the diffusion effect over the Auger recombination can be derived from the ratio $D_a \Delta N / [\mathcal{R}_{\mathrm{laser}}^2 (C_n + C_p) \Delta N^3]$. It can be checked that for the Auger recombinations to be commensurate with the diffusion effect, the excess carrier concentration has to be approximately $10^{20}\,\mathrm{cm}^{-3}$. This is two orders of magnitude higher than all the excess carrier concentration involved in TP.

4.1.3.4 Heat Generation Rate

The discussion of the heat generation rate is fairly simple. Though it is a linear effect, both the expressions of the direct heating and recombination heat still involve the bandgap and could therefore generate harmonics. Further, Auger recombinations also appear in the recombination heat. However, we have shown in Sect. 4.1.3.1 that only the constant mode E_{g0} needs to be considered and that Auger generates negligible harmonics in a diffusion-limited situation. In other words, the Fourier coefficients of the direct heating and recombination heat can be written

$$[(h\nu_{\text{laser}} - E_g)G^{\text{laser}}]_0 \approx (1 - R_0)(h\nu_{\text{laser}} - E_{g0})/(h\nu_{\text{laser}})(\alpha^{\text{laser}}_{\text{BTB}_0}\mathcal{P}^0_{\text{laser}}/2) \tag{4.50a}$$

$$[E_g\text{Rec}]_0 \approx E_{g0}\text{Rec}_0 \tag{4.50b}$$

$$[(h\nu_{\text{laser}} - E_g)G^{\text{laser}}]_1 \approx (1 - R_0)(h\nu_{\text{laser}} - E_{g0})/(h\nu_{\text{laser}})(\alpha^{\text{laser}}_{\text{BTB}_0}\mathcal{P}^0_{\text{laser}}/4) \tag{4.50c}$$

$$[E_g\text{Rec}]_1 \approx E_{g0}\text{Rec}_1 \tag{4.50d}$$

$$[(h\nu_{\text{laser}} - E_g)G^{\text{laser}}]_j \approx 0 \tag{4.50e}$$

$$[E_g\text{Rec}]_j \approx 0 \text{ for } j = 2, 3, \ldots. \tag{4.50f}$$

In conclusion, the heat generation rate does not induce extra harmonics.

4.1.3.5 Heat Diffusion Rate

In Si, the heat conduction occurs via phonons and not carriers. In other words, the heat conductivity k_{th} is independent from the carrier concentration [24]. Further, the low magnitude of the considered temperature perturbations allows us to neglect the variations in k_{th} with temperature. As a consequence, the heat conductivity and, hence, the heat diffusivity $D_{th} = k_{th}/(\rho c_p)$ are constant. In other words, the heat diffusion rate is linear and does not need any further discussion.

4.1.4 Summary

Our case study has shown that our model Eq. (4.32a, 4.32b) can be strongly simplified. Yet, the conditions of this case study have been chosen very close to the situations encountered during a TP experiment. The typical excess carrier concentration reached during a TP experiment are indeed of the order of ΔN_{ref}. All the conclusions of our case study are therefore valid for TP. For the carrier equations, this has two critical implications. First, coupling between the carrier equations, albeit limited, exists mostly due to the nonlinear diffusion rate and the nonlinear Auger recombinations. Second, only the constant and fundamental modes of the carrier distribution need to be considered. Other harmonics are excited by the Auger recombinations but their magnitudes are negligible since Auger is not a dominant phenomenon. For the heat equation, the simplifications go even further. First, only the constant and fundamental modes are excited. Second, no coupling exists between the heat equations for the constant and fundamental modes.

Note that the addition of a second harmonic to our case-study excess carrier distribution (4.36) would have led to the same conclusions. This can be proven by contradiction. The nonlinear generation rate indeed excites only the constant and fundamental modes of the excess carrier distribution, independently from the assumed excess carrier distribution. As a result, the consideration of a second harmonic commensurate with the constant and fundamental modes would lead to the absurd situation of the existence of this large second harmonic without any physical phenomenon accounting for its generation (the second harmonic generated by Auger recombinations is small in a diffusion-limited regime).

In other words, the general equations to be solved for the carrier and heat transport are the following

$$0 = D_{a0}\nabla^2(\Delta N_0) + 2\Re[D_{a1}^*\nabla^2(\Delta N_1)] + G_0^{\text{pump}} + G_0^{\text{probe}} - \text{Rec}_0 \tag{4.51a}$$

$$\begin{aligned} i\omega\Delta N_1 &= D_{a0}\nabla^2(\Delta N_1) + D_{a1}\nabla^2(\Delta N_0) \\ &\quad + D_{a2}\nabla^2(\Delta N_1^*) + G_1^{\text{pump}} - \text{Rec}_1 \end{aligned} \tag{4.51b}$$

$$\begin{aligned} 0 &= D_{th}\nabla^2(\Delta T_0) + 1/(\rho c_p)[(h\nu_{\text{pump}} - E_{g0})G_0^{\text{pump}} \\ &\quad + (h\nu_{\text{probe}} - E_{g0})G_0^{\text{probe}} + E_{g0}\text{Rec}_0] \end{aligned} \tag{4.51c}$$

$$i\omega\Delta T_1 = D_{th}\nabla^2(\Delta T_1) + 1/(\rho c_p)[(h\nu_{\text{pump}} - E_{g0})G_1^{\text{pump}} + E_{g0}\text{Rec}_1], \tag{4.51d}$$

with the carrier generation rates given by the following Beer's law

$$\begin{aligned} G_0^{\text{probe}}(x,y,z) &= \alpha_{\text{BTB}_0}^{\text{probe}}(1 - R_0^{\text{probe}})\mathcal{P}_{\text{probe}}^0 \exp\left(-\frac{x^2+y^2}{\mathcal{R}_{\text{probe}}^2}\right) \\ &\quad \times \exp(-\alpha_{\text{BTB}_0}^{\text{probe}} z)/(h\nu_{\text{probe}}) \end{aligned} \tag{4.52a}$$

$$\begin{aligned} G_0^{\text{pump}}(x,y,z) &= \alpha_{\text{BTB}_0}^{\text{pump}}(1 - R_0^{\text{pump}})\frac{\mathcal{P}_{\text{pump}}^0}{2} \exp\left(-\frac{x^2+y^2}{\mathcal{R}_{\text{pump}}^2}\right) \\ &\quad \times \exp(-\alpha_{\text{BTB}_0}^{\text{pump}} z)/(h\nu_{\text{pump}}) \end{aligned} \tag{4.52b}$$

$$\begin{aligned} G_1^{\text{pump}}(x,y,z) &= \alpha_{\text{BTB}_0}^{\text{pump}}(1 - R_0^{\text{pump}})\frac{\mathcal{P}_{\text{pump}}^0}{4} \exp\left(-\frac{x^2+y^2}{\mathcal{R}_{\text{pump}}^2}\right) \\ &\quad \times \exp(-\alpha_{\text{BTB}_0}^{\text{pump}} z)/(h\nu_{\text{pump}}). \end{aligned} \tag{4.52c}$$

As already mentioned, in this thesis, we make use of Smith's band-to-band absorption model [18] and Schenk's BGN model [12] to calculate $\alpha_{\text{BTB}_0}^{\text{probe}}$ and $\alpha_{\text{BTB}_0}^{\text{pump}}$. The expressions of Rec_0 and Rec_1, calculated from Eqs. (4.44) and (4.48), can be found in Appendix A.7. As for D_{a0}, D_{a1} and D_{a2}, they are calculated from Eq. (4.20b) using Eqs. (4.18a, 4.18b) and (4.40a, 4.40b) and Klaassen's mobility model [21].

We propose to use a final assumption so as to further simplify the model equations. In TP, the constant probe laser also shines on the sample, leading to a constant mode of the excess carrier distribution ΔN_0 somewhat larger than the fundamental mode $|\Delta N_1|$. As a result, the terms $2\Re[D_{a1}^* \nabla^2(\Delta N_1)]$ and $D_{a2}\nabla^2(\Delta N_1^*)$ can be neglected in first order. Based on this simplification, it is interesting to note that the Eq. (4.51a) can be written

$$\nabla^2(\Delta N_0) = -\frac{1}{D_{a0}}\left[G_0^{\text{pump}} + G_0^{\text{probe}} - \text{Rec}_0\right]. \tag{4.53}$$

In conclusion, the system (4.51a–4.51d) can now be written

$$0 = D_{a0}\nabla^2(\Delta N_0) + G_0^{\text{pump}} + G_0^{\text{probe}} - \text{Rec}_0 \tag{4.54a}$$

$$i\omega\Delta N_1 = D_{a0}\nabla^2(\Delta N_1) - \frac{D_{a1}}{D_{a0}}\left[G_0^{\text{pump}} + G_0^{\text{probe}} - \text{Rec}_0\right] + G_1^{\text{pump}} - \text{Rec}_1 \tag{4.54b}$$

$$0 = D_{th}\nabla^2(\Delta T_0) + 1/(\rho c_p)[(h\nu_{\text{pump}} - E_{g0})G_0^{\text{pump}} + (h\nu_{\text{probe}} - E_{g0})G_0^{\text{probe}} + E_{g0}\text{Rec}_0] \tag{4.54c}$$

$$i\omega\Delta T_1 = D_{th}\nabla^2(\Delta T_1) + 1/(\rho c_p)[(h\nu_{\text{pump}} - E_{g0})G_1^{\text{pump}} + E_{g0}\text{Rec}_1], \tag{4.54d}$$

where mainly the recombination terms Rec_0 and Rec_1 as well as D_{a1} still couple the carrier equations through their dependence upon the constant and fundamental modes of the excess carrier distribution. Notice that, since D_{a1} is negative in most cases (Fig. 4.6c), the term $\left[-D_{a1}/D_{a0}(G_0^{\text{pump}} + G_0^{\text{probe}} - \text{Rec}_0)\right]$ acts as a positive generation term, which will increase the fundamental mode of the excess carriers. This term is usually forgotten in most transport models for PMOR. For example, this term explains the difference between the steady-periodic and time-dependent calculations in [25]. It also explains the underestimation of the excess carriers in steady-periodic calculations observed in pages 165–166 of Ref. [13]. As we shall prove, it accounts for about 10 % of the fundamental mode of the excess carriers at high injection.

Equation (4.54a–4.54d) are our final model equations. In the next Sections, we look at the solutions of these equations for TP.

4.2 Solutions

In this Section, we look at the solutions of Eqs. (4.54a–4.54d). In order to build up our understanding of the behavior of these solutions, we propose to begin with simplified problems. In Sect. 4.2.1, we start by solving the one-dimensional linear problem when only the pump laser is shining. In spite of its quantitative incongruity, this one-dimensional solution offers qualitative understanding of the general behavior of the excess carriers and temperature with useful analytical expressions. Second, in Sect. 4.2.2, the linear problem is looked at on a three-dimensional sample. Since this problem can be solved analytically, we first discuss the analytical solution. We then compare the results of this analytical solution with numerical simulations run with FSEM [14] (Finite-element solver for SEMiconductors), a numerical transport simulator. The numerical results yielded by FSEM and the analytical solution will prove to be in excellent agreement. This gives us the sufficient confidence to, eventually, rely on FSEM to solve numerically the complete nonlinear three-dimensional problem of Eq. (4.54), the solutions of which are discussed in Sect. 4.2.3.

All the solutions below are discussed specifically for the TP630XP tool, i.e.:

- a 790 nm (1.57 eV) pump laser with 13.5 mW power modulated at a frequency $\omega = 1$ MHz and focused onto a radius $\mathcal{R}_{\text{pump}} = 0.5\,\mu$m on the sample
- when relevant (i.e. for the nonlinear problem), a 670 nm (1.85 eV) probe laser with 2.5 mW power focused onto a radius $\mathcal{R}_{\text{probe}} = 0.5\,\mu$m on the sample.

4.2.1 One-Dimensional Linear Solution

To understand the physics underlying Eqs. (4.54a–4.54d), let us start by solving the linear problem of a pump laser shining at $x = 0$ on a one-dimensional semi-infinite silicon sample in the lateral direction. By linear problem, we mean that the recombination rate is assumed to vary linearly with the excess carrier concentration, i.e. $\text{Rec}_j = \Delta N_j/\tau$ $(j = 0, 1)$ where τ is the carrier recombination lifetime. The linearity also implies that the ambipolar diffusivity is assumed independent from excess carrier concentration. In particular, we take $D_a = 8\,\text{cm}^2\,\text{s}^{-1}$, which is a typical value for a carrier injection of 10^{18} cm^{-3} in a lowly doped Si sample (Fig. 4.5). Finally, as an implicit result of the linearity, all coupling between the constant and fundamental modes of the carrier concentration vanishes.

Since the problem only has one lateral dimension, we need to consider that all carriers are generated at the surface, i.e. we need to integrate Eqs. (4.52b) and (4.52c) over depth and consider these generation rates as boundary conditions. Further, we assume a pump laser with zero lateral extension ($\mathcal{R}_{\text{pump}} = 0$). All the photon flux is therefore concentrated at $x = 0$. Following Eqs. (4.52b) and (4.52c), this gives respectively for the constant and fundamental modes of the surface carrier generation rate $G_0^{\text{pump}} = (1 - R_0^{\text{pump}})\mathcal{P}^0_{\text{pump}}/(2h\nu_{\text{pump}})$ and $G_1^{\text{pump}} = G_0^{\text{pump}}/2$. In summary, we need to solve the following four equations

$$\frac{\partial^2 \Delta N_0}{\partial x^2} - \frac{1}{L_{pl}^2}\Delta N_0 = 0 \tag{4.55a}$$

$$\frac{\partial^2 \Delta N_1}{\partial x^2} - \sigma_{pl}^2 \Delta N_1 = 0 \tag{4.55b}$$

$$\frac{\partial^2 \Delta T_0}{\partial x^2} + \frac{E_g}{k_{th}} \frac{\Delta N_0}{\tau} = 0 \tag{4.55c}$$

$$\frac{\partial^2 \Delta T_1}{\partial x^2} - \sigma_{th}^2 \Delta T_1 + \frac{E_g}{k_{th}} \frac{\Delta N_1}{\tau} = 0, \tag{4.55d}$$

where $\sigma_{pl} = \sqrt{1 + i\omega\tau}/L_{pl}$ and $\sigma_{th} = (1 + i)/L_{th}$ are respectively the wave vectors of the excess carrier and temperature distributions, with $L_{pl} = \sqrt{D_a\tau}$ and $L_{th} = \sqrt{2D_{th}/\omega}$ their respective low-frequency diffusion lengths.

The following Neumann boundary conditions need to be considered at $x = 0$ (under the pump beam)

$$-D_a \frac{\partial \Delta N_0}{\partial x}\bigg|_{x=0} = G_0^{\text{pump}} \tag{4.56a}$$

$$-D_a \frac{\partial \Delta N_1}{\partial x}\bigg|_{x=0} = \frac{G_0^{\text{pump}}}{2} \tag{4.56b}$$

$$-k_{th} \frac{\partial \Delta T_0}{\partial x}\bigg|_{x=0} = G_0^{\text{pump}}(h\nu_{\text{pump}} - E_g) \tag{4.56c}$$

$$-k_{th} \frac{\partial \Delta T_1}{\partial x}\bigg|_{x=0} = \frac{G_0^{\text{pump}}}{2}(h\nu_{\text{pump}} - E_g). \tag{4.56d}$$

As mentioned above, since we have surface absorption, the four non-homogeneous Neumann boundary conditions contain the information about the pump carrier generation and the direct heat generation.

Notice that the one-dimensional steady heat Eq. (4.55c) together with the boundary condition (4.56c) is an ill-posed problem. One-dimensional Laplace equations such as Eq. (4.55c) only accept linear solutions, i.e. solutions of the type $\Delta T_0 = Ax + B$, where A and B are two constants. Yet, this type of solution is incompatible with our generation terms. For the same reason, the one-dimensional steady ambipolar diffusion Eq. (4.55a) is also problematic when the carrier lifetime becomes increasingly long. In other words, this one-dimensional case study is not appropriate for the steady equations. It is, however, very instructive when it comes to the solutions to Helmholtz equations such as Eqs. (4.55b) and (4.55d). We will therefore here only study the case of the fundamental modes of the excess carrier and temperature distributions. From the behavior of these distributions, however, we will derive the expected (and confirmed in the next Sections) behavior of the constant modes.[7]

Solving Eqs. (4.55b) and (4.55d) with the boundary conditions (4.56b) and (4.56d), the fundamental modes of the excess carrier and temperature distributions

[7] Physically, the difference between the equations for the constant and fundamental modes can be understood from the mean of their respective generation terms. The mean of the harmonic generation term is zero, which signifies that the equations for the fundamental modes of the carriers and temperature do not need a sink term. Similarly, when the recombination lifetime is short enough, the recombination term serves as sink term for Eq. (4.55a), which explains why the problem is well posed in the short lifetime regime. As for Eq. (4.55c), it has no sink term; the problem is therefore always ill-posed.

read

$$\Delta N_1(x) = \frac{G_0^{\text{pump}}}{2D_a\sigma_{pl}} \exp(-\sigma_{pl}x) \tag{4.57a}$$

$$\Delta T_1(x) = \frac{G_0^{\text{pump}}}{2k_{th}} \Bigg[\frac{(h\nu_{\text{pump}} - E_g)}{\sigma_{th}} \exp(-\sigma_{th}x) + \frac{E_g}{L_{pl}^2(\sigma_{th}^2 - \sigma_{pl}^2)} \left(\frac{\exp(-\sigma_{pl}x)}{\sigma_{pl}} - \frac{\exp(-\sigma_{th}x)}{\sigma_{th}} \right) \Bigg]. \tag{4.57b}$$

All the qualitative physics concerning the fundamental modes of the excess carrier and temperature distributions, i.e. the *plasma* and *thermal waves*, is contained in formulas (4.57a, 4.57b). It is clearly seen that both behave as *damped waves* or *diffusion wave fields* (thoroughly described and studied in Ref. [17]), i.e. their lateral behavior shows an exponential decay superimposed with a harmonic behavior. As a consequence, throughout this thesis, we use the following description for the lateral behavior of ΔN_1 and ΔT_1

$$\Delta N_1(x) = |\Delta N_1(x=0)| \exp(-i\phi_{pl1}(x=0)) \exp(-x/L_{d1}^{pl}) \exp(-2i\pi x/\Lambda_1^{pl}) \tag{4.58a}$$

$$\Delta T_1(x) = |\Delta T_1(x=0)| \exp(-i\phi_{th1}(x=0)) \exp(-x/L_{d1}^{th}) \exp(-2i\pi x/\Lambda_1^{th}), \tag{4.58b}$$

where $|\Delta N_1(x=0)|$ [resp. $|\Delta T_1(x=0)|$], $\phi_{pl1}(x=0)$ [resp. $\phi_{th1}(x=0)$], L_{d1}^{pl} (resp. L_{d1}^{th}) and Λ_1^{pl} (resp. Λ_1^{th}) are respectively the four characteristics of the plasma (resp. thermal) wave, i.e. its amplitude and phase under the pump beam, its lateral decay length and its wavelength.

Figures 4.10a–d respectively show the behavior as a function of the inverse carrier recombination lifetime ($1/\tau$) of the amplitudes, decay lengths, phases and wavelengths of the one-dimensional plasma and thermal waves of Eqs. (4.57a, 4.57b). The sample is undoped Si, so that E_g is 1.12 eV, $\rho = 2.3\ 10^{-3}\,\text{kg}\cdot\text{cm}^{-3}$, $c_p = 700\,\text{J}\cdot\text{kg}^{-1}\,\text{K}^{-1}$ and $k_{th} = 1.3\,\text{W}\cdot\text{cm}^{-1}\,\text{K}^{-1}$. Two different regimes can be identified. First, in the short-lifetime regime, or *recombination-limited* regime (corresponding to highly doped or highly injected Si), only the thermal wave shows a wave-like behavior. The plasma wave is a decaying exponential (zero phase and infinite wavelength). Second, in the long-lifetime regime, or *diffusion-limited* regime, the plasma and thermal waves have very similar wave-like behaviors. The asymptotical behaviors of the presented parameters at short and long lifetimes are also given in Fig. 4.10.

Notice already that the orders of magnitude appearing in Fig. 4.10 are linked to the one-dimensional character of this example and to the infinitely small pump radius.[8] As shown in the next examples, the amplitude of both the plasma and the

[8] These solutions are actually Green functions, see next Section

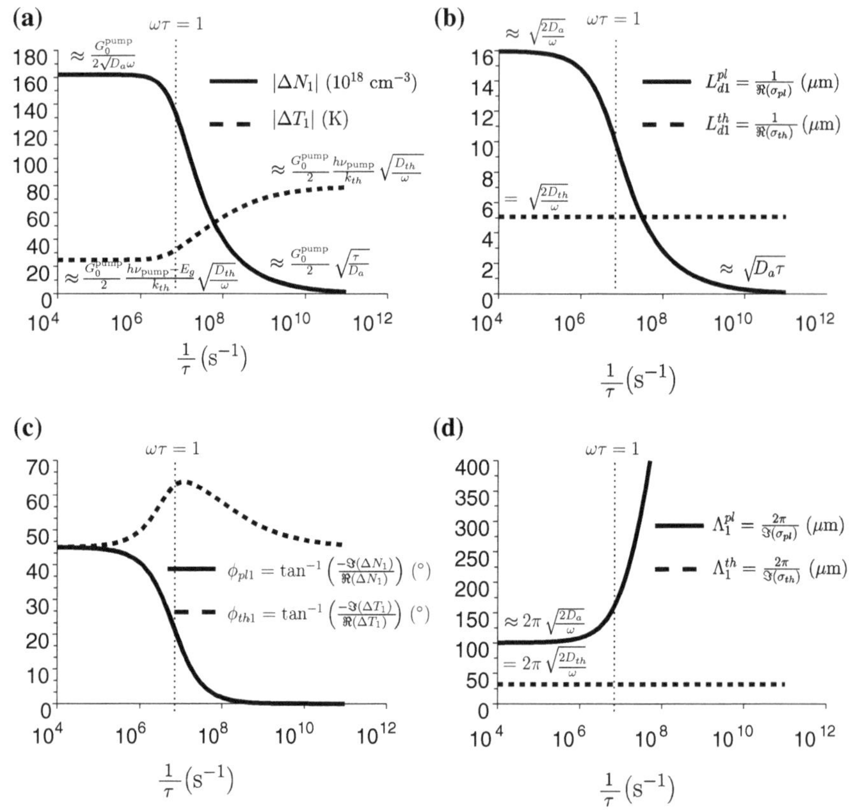

Fig. 4.10 Variation with recombination lifetime in the *one-dimensional* plasma wave ΔN_1 (*full lines*) and thermal wave ΔT_1 (*dashed lines*), solutions of Eqs. (4.55a)–(4.55d) in the case of a TP pump laser. **a** The amplitude and **c** phase under the beam ($x = 0$) as well as **b** the lateral decay length and **d** lateral wavelength are derived from the analytical solutions expressed in formulas (4.57a, 4.57b). Interesting asymptotic behaviors are also shown. The *vertical line* corresponds to $\omega\tau = 1$, i.e. the limit between the diffusion-limited regime (*left*, long lifetime) and the recombination-limited regime (*right*, short lifetime)

thermal waves is strongly reduced in a three-dimensional geometry with a laterally distributed pump irradiance (about two to three orders of magnitude less). This can partly be attributed to the extra two degrees of freedom for diffusion. As for the phase, it is closely related to the type of absorption [26] (surface-restricted or not). It therefore also changes when changing dimensionality. Finally, the diffusion lengths and wavelengths are also reduced in a three-dimensional problem. However, the variations are less than one order of magnitude.

The lateral behavior of the plasma and thermal waves of Eqs. (4.58a, 4.58b) is of the utmost importance in this work since it explains the final behavior of the ΔR_{ac} offset curves in Chap. 6. For a better visualization of this lateral behavior, we therefore represent respectively in Fig. 4.11a and b the normalized amplitude and the phase of the one-dimensional plasma and thermal waves in the case of a long recombination lifetime ($\tau = 10^{-4}$ s). The separate representation of the lateral behavior of the amplitude and phase of these damped waves in Fig. 4.11a and b, which we call the *complex* representation, is very convenient since it splits the damped wave into its damping behavior (amplitude, decay length) and its wave-like behavior (phase, wavelength). As highlighted in Fig. 4.11a and b, the slopes of respectively the amplitude and phase behavior give a direct access to respectively the decay length and the wavelength of the damped waves. For this reason, this representation is widely used in this work.

For completeness, Fig. 4.11c shows the more common *real* representation of the normalized plasma and thermal waves, i.e. the lateral behavior of $2\Re[\Delta N_1(x)/|\Delta N_1(x=0)|]$ and $2\Re[\Delta T_1(x)/|\Delta T_1(x=0)|]$. It can clearly be observed that the characteristics of both damped waves become difficult to identify. As a consequence, this representation is not used in this work.

As a final remark, note that, though only the fundamental modes of the excess carrier and temperature distributions have been calculated in this Section, it can be intuitively expected (and confirmed in the next Sections) that the constant modes ΔN_0 and ΔT_0 present a similar behavior, except that they have zero phase and infinite wavelength (i.e. they are real). They are therefore written

$$\Delta N_0(x) = \Delta N_0(x=0)\exp(-x/L_{d0}^{pl}) \tag{4.59a}$$

$$\Delta T_0(x) = \Delta T_0(x=0)\exp(-x/L_{d0}^{th}), \tag{4.59b}$$

where $\Delta N_0(x=0)$ [resp. $\Delta T_0(x=0)$] and L_{d0}^{pl} (resp. L_{d0}^{th}) are respectively the amplitude and lateral decay length of the constant carrier (resp. temperature) distribution. Note that the constant modes do not show a wave-like behavior. We will therefore, without any ambiguity, often refer to the fundamental modes of the carrier and temperature distributions as respectively the plasma and thermal waves. Equations (4.58a, 4.58b) and (4.59a, 4.59b) are critical for the general understanding of the carrier and temperature distributions and, as we shall see, of the AC reflectance.[9]

4.2.2 Three-Dimensional Linear Solution

Since the one-dimensional linear problem results in ill-posed steady equations and misquantified fundamental modes, it is necessary to investigate the problem of a

---

9 It is obvious that these definitions do not hold in all conditions. For instance, the thermal wave of Eq. (4.57b), even in the one-dimensional case, is the superposition of two damped waves and therefore can only be rigorously written as Eq. (4.58b) when one of the two is dominant. Similarly, when the pump irradiance distribution is not considered to be strictly punctual, the exponential decay is perturbed (the curvature becomes positive under the pump beam). However, these definitions still prove to be very convenient for the overall understanding of the lateral behavior of plasma and thermal waves as well as of the AC reflectance.

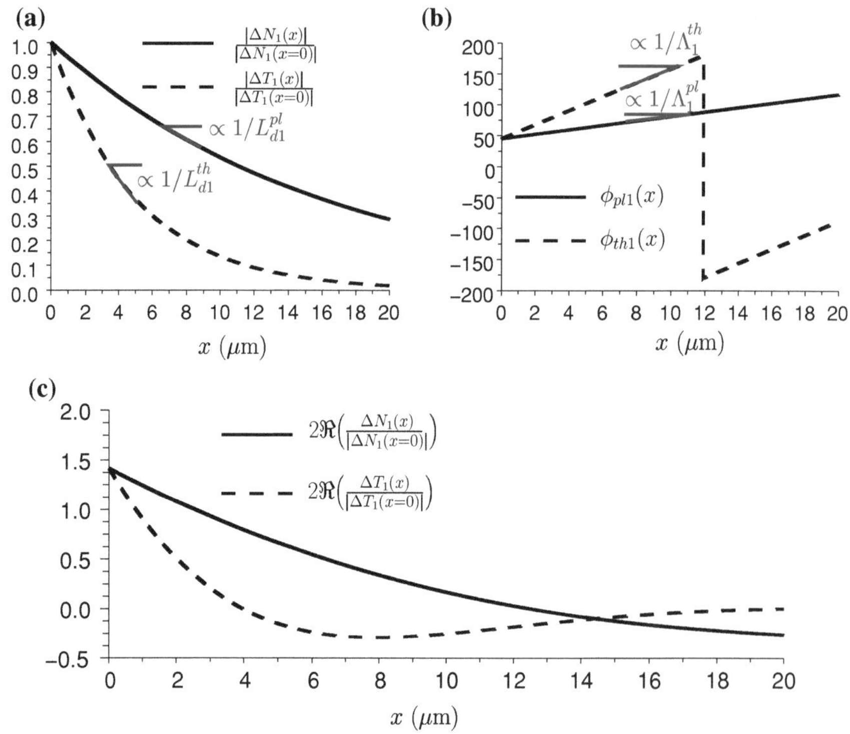

Fig. 4.11 *Complex* and *real* representations of lateral behavior of the normalized plasma and thermal waves for $\tau = 10^{-3}$ s. Complex representation: variation with the lateral position x in **a** the normalized amplitude and **b** the phase of the plasma (*full lines*) and thermal (*dashed lines*) waves. This representation separates the damping and wave-like behavior of each wave. **c** Real representation: variation with the lateral position x in the (*doubled*) real values of the plasma (*full line*) and thermal (*dashed line*) waves. The damping and wave-like behaviors are simultaneously shown and therefore difficult to discriminate

pump laser shining on a three-dimensional sample. In other words, we want to solve the following equations

$$\nabla^2(\Delta N_0) - \frac{1}{L_d^2}\Delta N_0 = -\frac{1}{D_a}G_0^{\text{pump}} \tag{4.60a}$$

$$\nabla^2(\Delta N_1) - \sigma_{pl}^2\Delta N_1 = -\frac{1}{D_a}G_1^{\text{pump}} \tag{4.60b}$$

$$\nabla^2(\Delta T_0) = -\frac{1}{k_{th}}[(h\nu_{\text{pump}} - E_g)G_0^{\text{pump}} + E_g\frac{\Delta N_0}{\tau}] \tag{4.60c}$$

$$\nabla^2(\Delta T_1) - \sigma_{th}^2\Delta T_1 = -\frac{1}{k_{th}}[(h\nu_{\text{pump}} - E_g)G_1^{\text{pump}} + E_g\frac{\Delta N_1}{\tau}], \tag{4.60d}$$

where G_0^{pump} and G_1^{pump} follow respectively Eqs. (4.52b) and (4.52c). Note that the problem summarized in Eqs. (4.60a–4.60d) and (4.61a–4.61d) has cylindrical symmetry around the pump beam [situated at $(x, y) = (0, 0)$]. Equations (4.60a–4.60d) are to be considered with the following homogeneous Neumann boundary conditions at the top surface $z = 0$

$$-D_a \frac{\partial \Delta N_0}{\partial z}\bigg|_{z=0} = 0 \tag{4.61a}$$

$$-D_a \frac{\partial \Delta N_1}{\partial z}\bigg|_{z=0} = 0 \tag{4.61b}$$

$$-k_{th} \frac{\partial \Delta T_0}{\partial z}\bigg|_{z=0} = 0 \tag{4.61c}$$

$$-k_{th} \frac{\partial \Delta T_1}{\partial z}\bigg|_{z=0} = 0, \tag{4.61d}$$

and at $r = 0$ (imposed by cylindrical coordinates)

$$-D_a \frac{\partial \Delta N_0}{\partial r}\bigg|_{r=0} = 0 \tag{4.62a}$$

$$-D_a \frac{\partial \Delta N_1}{\partial r}\bigg|_{r=0} = 0 \tag{4.62b}$$

$$-k_{th} \frac{\partial \Delta T_0}{\partial r}\bigg|_{r=0} = 0 \tag{4.62c}$$

$$-k_{th} \frac{\partial \Delta T_1}{\partial r}\bigg|_{r=0} = 0, \tag{4.62d}$$

where $r = \sqrt{x^2 + y^2}$.

This problem can be solved analytically. The analytical solution is derived in Sect. 4.2.2.1 assuming a semi-infinite sample. The problem is also solved numerically in Sect. 4.2.2.2 using the following extra boundary condition

$$-D_a \frac{\partial \Delta N_0}{\partial z}\bigg|_{z=L_z} = 0 \tag{4.63a}$$

$$-D_a \frac{\partial \Delta N_1}{\partial z}\bigg|_{z=L_z} = 0 \tag{4.63b}$$

$$\Delta T_0\bigg|_{z=L_z} = 0 \tag{4.63c}$$

$$\Delta T_1\bigg|_{z=L_z} = 0, \tag{4.63d}$$

where L_z is the total thickness of the sample.

We will see that the two, analytical and numerical, approaches are in excellent agreement (<1 % error).

4.2.2.1 An Analytical Approach

The problem summarized in Eqs. (4.60a–4.60d), (4.61a–4.61d) and (4.62a–4.62d) can be solved by combining the Green function [27] and Hankel transform [28] theories. The derivation of these analytical solutions is a three-step procedure. First, we seek the Green functions $g^{N_0}(r, r', z, z')$, $g^{N_1}(r, r', z, z')$, $g^{T_0}(r, r', z, z')$ and $g^{T_1}(r, r', z, z')$ of Eqs. (4.60a–4.60d) respectively, i.e their solutions for a Dirac peak generation term $\delta(r - r', z - z')$ located at position (r', z'). In other words, we have to solve the following problem

$$\left[\frac{\partial^2}{\partial z^2} + \frac{1}{r}\frac{\partial}{\partial r}\left(r\frac{\partial}{\partial r}\right) - \frac{1}{L_d^2}\right] g^{N_0}(r, r', z, z') = -\frac{1}{D_a}\delta(r - r', z - z') \quad (4.64a)$$

$$\left[\frac{\partial^2}{\partial z^2} + \frac{1}{r}\frac{\partial}{\partial r}\left(r\frac{\partial}{\partial r}\right) - \sigma_{pl}^2\right] g^{N_1}(r, r', z, z') = -\frac{1}{D_a}\delta(r - r', z - z') \quad (4.64b)$$

$$\left[\frac{\partial^2}{\partial z^2} + \frac{1}{r}\frac{\partial}{\partial r}\left(r\frac{\partial}{\partial r}\right)\right] g^{T_0}(r, r', z, z') = -\frac{1}{D_{th}}\delta(r - r', z - z') \quad (4.64c)$$

$$\left[\frac{\partial^2}{\partial z^2} + \frac{1}{r}\frac{\partial}{\partial r}\left(r\frac{\partial}{\partial r}\right) - \sigma_{th}^2\right] g^{T_1}(r, r', z, z') = -\frac{1}{D_{th}}\delta(r - r', z - z'). \quad (4.64d)$$

Second, we write the Hankel transforms of Eqs. (4.64a–4.64d) and their boundary conditions. This yields

$$\left[\frac{\partial^2}{\partial z^2} - \left(\xi^2 + \frac{1}{L_{pl}^2}\right)\right] \mathcal{G}^{N_0}(\xi, r', z, z') = -\frac{J_0(\xi r')}{2\pi D_a}\delta_N(z - z') \quad (4.65a)$$

$$\left[\frac{\partial^2}{\partial z^2} - \underbrace{(\xi^2 + \sigma_{pl}^2)}_{=\mathcal{S}_{pl}^2(\xi)}\right] \mathcal{G}^{N_1}(\xi, r', z, z') = -\frac{J_0(\xi r')}{2\pi D_a}\delta_N(z - z') \quad (4.65b)$$

$$\left[\frac{\partial^2}{\partial z^2} - \xi^2\right] \mathcal{G}^{T_0}(\xi, r', z, z') = -\frac{J_0(\xi r')}{2\pi D_{th}}\delta_T(z - z') \quad (4.65c)$$

$$\left[\frac{\partial^2}{\partial z^2} - \underbrace{(\xi^2 + \sigma_{th}^2)}_{=\mathcal{S}_{th}^2(\xi)}\right] \mathcal{G}^{T_1}(\xi, r', z, z') = -\frac{J_0(\xi r')}{2\pi D_{th}}\delta_T(z - z'), \quad (4.65d)$$

where $J_0(\xi r')$ is the Bessel function of first kind of order zero [29] and $\mathcal{G}^{N_j}(\xi, r', z, z')$ and $\mathcal{G}^{T_j}(\xi, r', z, z')$ are the Hankel transforms of respectively $g^{N_j}(r, r', z, z')$ and $g^{T_j}(r, r', z, z')$ $(j = 0, 1)$, i.e.

$$\mathcal{G}^{N_j}(\xi, r', z, z') = \int_0^{+\infty} g^{N_j}(r, r', z, z') J_0(\xi r) r dr \tag{4.66a}$$

$$\mathcal{G}^{T_j}(\xi, r', z, z') = \int_0^{+\infty} g^{T_j}(r, r', z, z') J_0(\xi r) r dr. \tag{4.66b}$$

The reason for moving to the Hankel domain is obvious from the shape of Eqs. (4.65a–4.65d), where we can see that the partial-derivative equations have been turned into differential equations, i.e. the radial derivatives now appear as the algebraic term ξ^2. These are direct applications of the properties of Hankel transforms [28]. These equations are now ordinary second-order differential equations similar to their one-dimensional counterparts, Eqs. (4.55a–4.55d). However, one important difference is that Eqs. (4.65a–4.65d) are all Helmholtz equations, even at zero frequency (presence of the ξ^2 in the coefficient before the Green function). Since no Laplace equations are found here, the three-dimensional ΔN_0 and ΔT_0 do exist. Physically, this can be understood from the fact that, in three dimensions, the extra dimensions can be used as sink term (this is what mathematically represents the ξ^2 term). In particular, since ΔN_0 and ΔT_0 can readily be obtained from the solutions of the fundamental mode by setting a vanishing frequency and the correct laser irradiance, we focus on the latter modes for the following calculations.

Written in the Hankel domain, the boundary conditions (4.61a–4.61d) read

$$-D_a \frac{\partial \mathcal{G}^{N_1}(\xi, r', z, z')}{\partial z}|_{z=0} = 0 \tag{4.67a}$$

$$-k_{th} \frac{\partial \mathcal{G}^{N_1}(\xi, r', z, z')}{\partial z}|_{z=0} = 0. \tag{4.67b}$$

Extra boundary conditions are also needed at the position $z = z'$ of the source term, which are provided by the following properties of the Green Function [17]

$$\mathcal{G}^{N_1}(\xi, r', z, z')|_{z=(z')^+} - \mathcal{G}^{N_1}(\xi, r', z, z')|_{z=(z')^-} = 0 \tag{4.68a}$$

$$\frac{\partial \mathcal{G}^{N_1}(\xi, r', z, z')}{\partial z}|_{z=(z')^+} - \frac{\partial \mathcal{G}^{N_1}(\xi, r', z, z')}{\partial z}|_{z=(z')^-} = -\frac{J_0(\xi r')}{2\pi D_a} \tag{4.68b}$$

$$\mathcal{G}^{T_1}(\xi, r', z, z')|_{z=(z')^+} - \mathcal{G}^{T_1}(\xi, r', z, z')|_{z=(z')^-} = 0 \tag{4.68c}$$

$$\frac{\partial \mathcal{G}^{T_1}(\xi, r', z, z')}{\partial z}|_{z=(z')^+} - \frac{\partial \mathcal{G}^{T_1}(\xi, r', z, z')}{\partial z}|_{z=(z')^-} = -\frac{J_0(\xi r')}{2\pi D_{th}}, \tag{4.68d}$$

where the $(z')^+$ (resp. $(z')^-$) is a value of z infinitely close to z' but larger (resp. lower). The discontinuity of the first derivative of the Green function can be demonstrated by integration of Eqs. (4.65a–4.65d) (pages 23–24 of Ref. [17]).

The solutions to Eqs. (4.65b) and (4.65d) with the conditions (4.67a, 4.67b) and (4.68a–4.68d) are

$$\mathcal{G}^{N_1}(\xi, r', z, z') = \frac{J_0(\xi r')}{4\pi D_a \mathcal{S}_{pl}(\xi)} \left\{ \exp[-\mathcal{S}_{pl}(\xi)|z - z'|] + \exp[-\mathcal{S}_{pl}(\xi)(z + z')] \right\} \tag{4.69a}$$

$$\mathcal{G}^{T_1}(\xi, r', z, z') = \frac{J_0(\xi r')}{4\pi D_{th} \mathcal{S}_{th}(\xi)} \left\{ \exp[-\mathcal{S}_{th}(\xi)|z - z'|] + \exp[-\mathcal{S}_{th}(\xi)(z + z')] \right\}. \tag{4.69b}$$

Third, though algebraically tedious, the Hankel transform $\mathcal{N}_1(\xi, z)$ of the excess carrier distribution can now readily be obtained from its Green function (Eqs. (4.69a, 4.69b)), following Green's theorem [27], i.e. one needs to convolute the Green function with its generation term, i.e.

$$\begin{aligned} \mathcal{N}_1(\xi, z) &= 2\pi \int_0^{+\infty} r' dr' \int_0^{+\infty} dz' \mathcal{G}^{N_1}(\xi, r', z, z') G_1^{\text{pump}}(r', z') \\ &= \frac{\alpha_{BTB}^{\text{pump}}(1 - R_0^{\text{pump}})\mathcal{P}_{\text{pump}}^0 \mathcal{R}_{\text{pump}}^2}{4D_a h\nu_{\text{pump}}} \frac{\exp(-\xi^2 \mathcal{R}_{\text{pump}}^2/4)}{(\alpha_{BTB}^{\text{pump}})^2 - \mathcal{S}_{pl}^2(\xi)} \\ &\quad \times \left\{ \frac{\alpha_{BTB}^{\text{pump}}}{\mathcal{S}_{pl}(\xi)} \exp[-\mathcal{S}_{pl}(\xi) z] - \exp(-\alpha_{BTB}^{\text{pump}} z) \right\}, \end{aligned} \tag{4.70}$$

Similarly, we have for the Hankel transform $\mathcal{T}_1(\xi, z)$ of the temperature distribution

$$\begin{aligned} \mathcal{T}_1(\xi, z) &= \frac{2\pi}{\rho c_p} \int_0^{+\infty} r' dr' \int_0^{+\infty} dz' \mathcal{G}^{T_1}(\xi, r', z, z') \\ &\quad \left[(h\nu_{\text{pump}} - E_g) G_1^{\text{pump}}(r', z') + E_g \frac{\Delta N_1(r', z')}{\tau} \right] \\ &= \frac{\alpha_{BTB}^{\text{pump}}(1 - R_0^{\text{pump}})\mathcal{P}_{\text{pump}}^0 \mathcal{R}_{\text{pump}}^2}{4k_{th} h\nu_{\text{pump}}} \exp(-\xi^2 \mathcal{R}_{\text{pump}}^2/4) \\ &\quad \times \left\{ \frac{(h\nu_{\text{pump}} - E_g)}{[(\alpha_{BTB}^{\text{pump}})^2 - \mathcal{S}_{th}^2(\xi)]} \left\{ \frac{\alpha_{BTB}^{\text{pump}}}{\mathcal{S}_{th}(\xi)} \exp[-\mathcal{S}_{th}(\xi) z] - \exp(-\alpha_{BTB}^{\text{pump}} z) \right\} \right. \\ &\quad + \left\{ \frac{E_g}{\tau D_a[(\alpha_{BTB}^{\text{pump}})^2 - \mathcal{S}_{pl}^2(\xi)]} \right. \\ &\quad \times \left\{ \frac{1}{\mathcal{S}_{th}^2(\xi) - (\alpha_{BTB}^{\text{pump}})^2} \left\{ \frac{\alpha_{BTB}^{\text{pump}}}{\mathcal{S}_{th}(\xi)} \exp[-\mathcal{S}_{th}(\xi) z] - \exp(-\alpha_{BTB}^{\text{pump}} z) \right\} \right. \\ &\quad \left. \left. \left. + \frac{\alpha}{\mathcal{S}_{pl}(\xi)} \frac{1}{\mathcal{S}_{pl}^2(\xi) - \mathcal{S}_{th}^2(\xi)} \left\{ \frac{\mathcal{S}_{pl}(\xi)}{\mathcal{S}_{th}(\xi)} \exp[-\mathcal{S}_{th}(\xi) z] - \exp(-\mathcal{S}_{pl}(\xi) z) \right\} \right\} \right\} \right\}, \end{aligned} \tag{4.71}$$

From these analytical Hankel transforms, it is then easy to derive the excess carrier and temperature distributions by inverse Hankel tranform, i.e.

$$\Delta N_1(r,z) = \int_0^{+\infty} \mathcal{N}_1(\xi,z) J_0(\xi r) \xi d\xi \tag{4.72a}$$

$$\Delta T_1(r,z) = \int_0^{+\infty} \mathcal{T}_1(\xi,z) J_0(\xi r) \xi d\xi. \tag{4.72b}$$

These integrals being not trivial, the derived analytical solution is mostly convenient if one needs the values of the excess carrier and temperature distributions at a few positions only. Further, owing to the complexity of these expressions, it is difficult to use them as such for further understanding of the behavior of these distributions. We therefore characterize the lateral behavior of ΔN_0, ΔN_1, ΔT_0 and ΔT_1 as we have for the one-dimensional case, i.e. using Eqs. (4.58a, 4.58b) and (4.59a, 4.59b).

As can be seen in Fig. 4.12, the qualitative behavior of ΔN_1 and ΔT_1 resembles much the one-dimensional case shown in Fig. 4.10. The discussion of the latter figure is therefore still relevant. Particularly, it can be seen that the transition between the diffusion-limited and the recombination-limited regimes also occurs around $1/\tau = 10^7\ \mathrm{s}^{-1}$, i.e. when $\omega\tau \approx 1$. Nevertheless, Fig. 4.12 also presents some new interesting features. First, as already hinted at previously, the orders of magnitude are very different in the one-dimensional and three-dimensional cases. Second, solutions for the constant modes exist in a three-dimensional geometry. It can be seen that, as expected, the constant modes of both the carrier and temperature distributions show a general behavior very similar to the fundamental modes. However, three differences can be noticed. First, ΔN_0 and ΔT_0 always have zero phase and infinite wavelength (i.e. they are real). Second, their amplitude is always somewhat more than twice that of the fundamental mode. This is due both to the difference in excitation powers of the two modes (factor 2) and to transient effects. At low modulation frequency (static mode), the ratio of their amplitude would be exactly equal to 2. Finally, the decay lengths of the constant modes are longer than those of the fundamental modes. Finally, note that the appearance of a bump in the thermal decay length L_{d1}^{th} and Λ_1^{th} in the $10^{-7} - 10^{-10}$ s range is a consequence of the recombination heat. Figure 4.10b and d indeed show the behavior of $1/\Re(\sigma_{th})$ and $2\pi/\Im(\sigma_{th})$, which only include the contribution of direct heating.

4.2.2.2 A Numerical Approach

As can be seen from the above section, solving analytically even the linear three-dimensional problem is far from being trivial. Yet, we have seen in this chapter that numerous nonlinear effects should be taken into account. Though these effects could partially be accounted for with limited error by plugging Eqs. (4.70), (4.71)

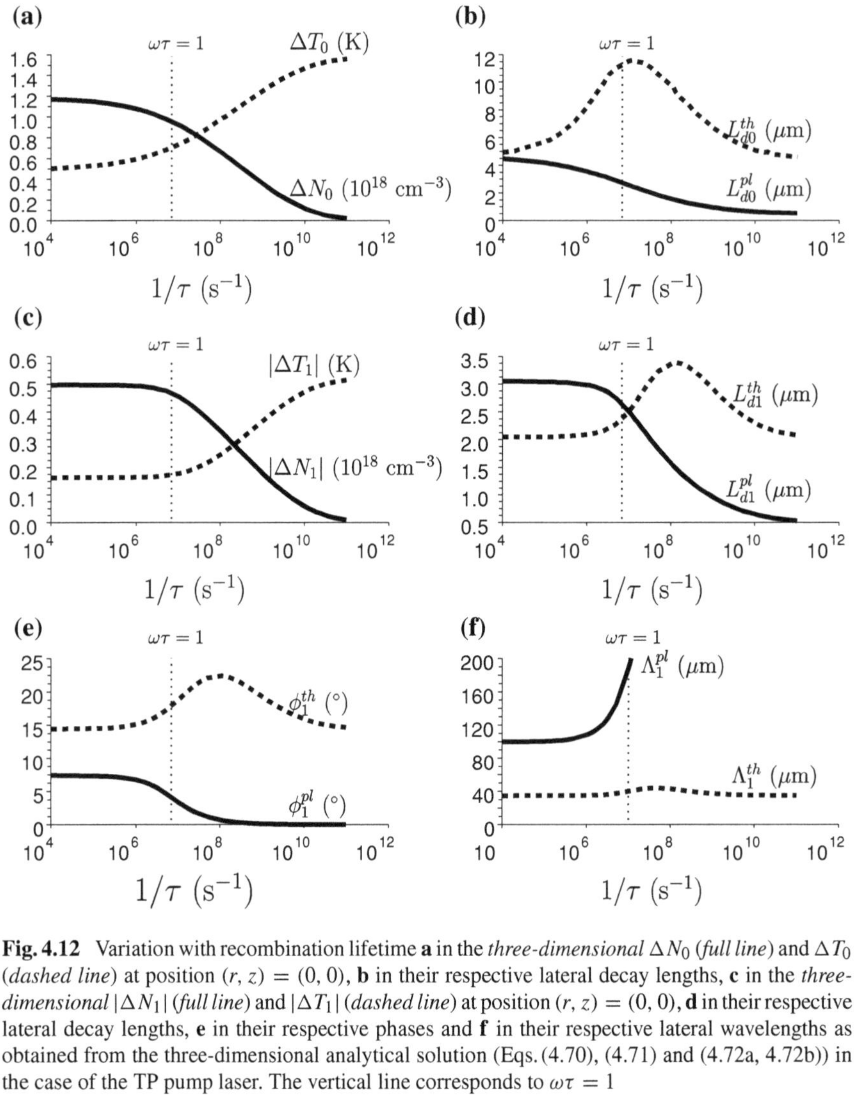

Fig. 4.12 Variation with recombination lifetime **a** in the *three-dimensional* ΔN_0 (*full line*) and ΔT_0 (*dashed line*) at position $(r, z) = (0, 0)$, **b** in their respective lateral decay lengths, **c** in the *three-dimensional* $|\Delta N_1|$ (*full line*) and $|\Delta T_1|$ (*dashed line*) at position $(r, z) = (0, 0)$, **d** in their respective lateral decay lengths, **e** in their respective phases and **f** in their respective lateral wavelengths as obtained from the three-dimensional analytical solution (Eqs. (4.70), (4.71) and (4.72a, 4.72b)) in the case of the TP pump laser. The vertical line corresponds to $\omega\tau = 1$

and (4.72a, 4.72b) into a nonlinear Newton loop [30], we still believe that a fully numerical approach is more appropriate.

Numerous arguments are in favor of a numerical approach. We mention here only the two most critical. First, the number of involved assumptions can be greatly reduced. The above analytical solution supposes homogeneous ambipolar diffusivity, bandgap and recombination lifetime. This proves to work with limited error for TP [31] but it serves against the versatility of the approach. It would therefore be interesting to have a resolution technique which can account for the space variations in these parameters. Second, the analytical solutions require the numerical calculation of two integrals for each position (r, z). This is quite convenient for the calculation of $\Delta N(r, z)$ at one single position but turns out to be too time-consuming if the values at multiple positions are sought. As a consequence, it is clear that the availability of a more general, e.g. numerical, technique would be helpful.

To solve carrier and heat transport problems in complex conditions, one generally resorts to commercially available numerical simulators such as Medici [32] or Sentaurus Device [33]. However, these transport simulators have been designed for the resolution of carrier and heat transport in typical conditions of modern nanoelectronics and therefore present unacceptable drawbacks. First, they solve the original transport Eqs. (4.6a–4.6c) and (4.8a, 4.8b), i.e. not their highly simplified versions summarized by Eqs. (4.54a–4.54d). This involves unacceptably time-consuming calculations. In particular, only time-dependent calculations are then possible (no simplified steady-periodic equations such as we derived in Sect. 4.1.3). Second, direct heating is not implemented in these commercial software packages. Since this is the dominant heating phenomenon at long recombination lifetimes, it can hardly ever be neglected.

Part of our investigations have therefore been dedicated to the development of a package to solve this problem by the finite-element method [14]. This project, called a Finite Element Solver for Semiconductors (FSEM) has initially been launched by F. Dortu during his Ph.D. [13]. In the framework of the present work, several months have been spent further developing the code (nonlinear steady drift-diffusion equations, nonlinear steady-periodic ambipolar equation,...) and deepening the understanding of the underlying physics. We have also used FSEM intensively.[10] The basic equations and implementation of FSEM being already described in great detail elsewhere [13], we here only focus on the powerful results FSEM can help obtain.

The three-dimensional linear problem of Eqs. (4.60a–4.60d) and (4.61a–4.61d) has been solved analytically and the solutions are already shown in Fig. 4.12. The comparison of these results with those yielded by FSEM therefore enables one to estimate the error made in our numerical calculations. It is further of great help when it comes to determining the optimal simulation settings for accurate but fast simulations (sample size, refinement of the mesh,...). We define the relative difference between the analytical and numerical solution as follows

$$\delta_{\%}\mathcal{V} = \left| \frac{\mathcal{V}_{\text{analytical}} - \mathcal{V}_{\text{numerical}}}{\mathcal{V}_{\text{analytical}}} \right|, \tag{4.73}$$

where $\mathcal{V}$ is an arbitrary calculated parameter, the value of which is $\mathcal{V}_{\text{analytical}}$ in the analytical solution and $\mathcal{V}_{\text{numerical}}$ in the numerical solution. Note that, since both the numerical and analytical solutions suffer from numerical errors, it is not possible to determine from which solution the error originates.

[10] I take the advantage of this Section to stress once again my deep gratitude towards Fabian Dortu, the initial developer of FSEM. His help and all the energy he put into FSEM have been the key to a lot of the successful results presented in this thesis.

We present in Fig. 4.13 the results obtained in the case of a sample with dimensions 600 μm both in the vertical and radial directions. First, it can be checked that the numerical and analytical solutions are in excellent quantitative agreement since the difference in all shown parameters is less than one percent (in most cases). Notice that the calculation of the various decay lengths and wavelengths (Figs. 4.13b, d and f) involve extra calculations and therefore some error propagation. This is the reason why the difference between the numerical and analytical values is slightly higher than for the values at position $(r, z) = (0, 0)$ (Figs. 4.13a, c and e). Further, the relative difference in plasma wavelength is also large at short lifetimes since the plasma wavelength itself diverges.

Based on this excellent agreement, we believe FSEM is an ideal basis to build further understanding upon. We therefore use it in the next Section to solve the full three-dimensional nonlinear problem, i.e. our initial problem.

4.2.3 Three-Dimensional Nonlinear Solution

Moving back to the original nonlinear problem of a modulated pump laser and a constant probe laser shining[11] on a three-dimensional homogeneous silicon sample is now quite a simple step. We know the expected behavior of ΔN_0, ΔN_1, ΔT_0 and ΔT_1 (Eqs. (4.58a, 4.58b) and (4.59a, 4.59b)). We have also shown that we have a reliable tool to solve the problem numerically (FSEM).

The transport equations to be solved here are summarized in Eqs. (4.54a–4.54d) with the boundary conditions of Eqs. (4.61a–4.61d), (4.62a–4.62d) and (4.63a–4.63d). The nonlinearity of the equations is accounted for via external models which have been introduced and discussed in Sect. 4.1.3.

The calculated variations in ΔN_j $(j = 0, 1)$ as a function of p-type doping concentration in the case of coincident probe and pump beams are shown in Fig. 4.14. Four levels of refinement of the model are presented. First, the black curves represent the behavior of the carrier distributions when no BGN is assumed. Second, the blue curves show the impact of the consideration of a BGN-dependent absorption coefficient. Third, the green curves illustrate the behavior of ΔN_j $(j = 0, 1)$ when BGN impacts both the absorption coefficient and the ambipolar diffusivity (via the counter-diffusive BGN effect). Finally, the red curves show the complete solution accounting also for the coupling due to the nonlinear diffusion rate. It can clearly be seen that BGN has a major impact on all plotted parameters, especially via its impact on the absorption coefficient.[12] As expected, the coupling due to the nonlinear diffusion rate only impacts the value of $|\Delta N_1|$ at low doping, where it accounts for about 10 % of its total amplitude. Though the obtained excess carrier distribution

[11] The probe laser has not been considered in the previous Sections since only linear systems have been discussed.

[12] Yet, as shown earlier, the model used to account for this effect typically underestimates α_{BTB} (Fig. 4.2)

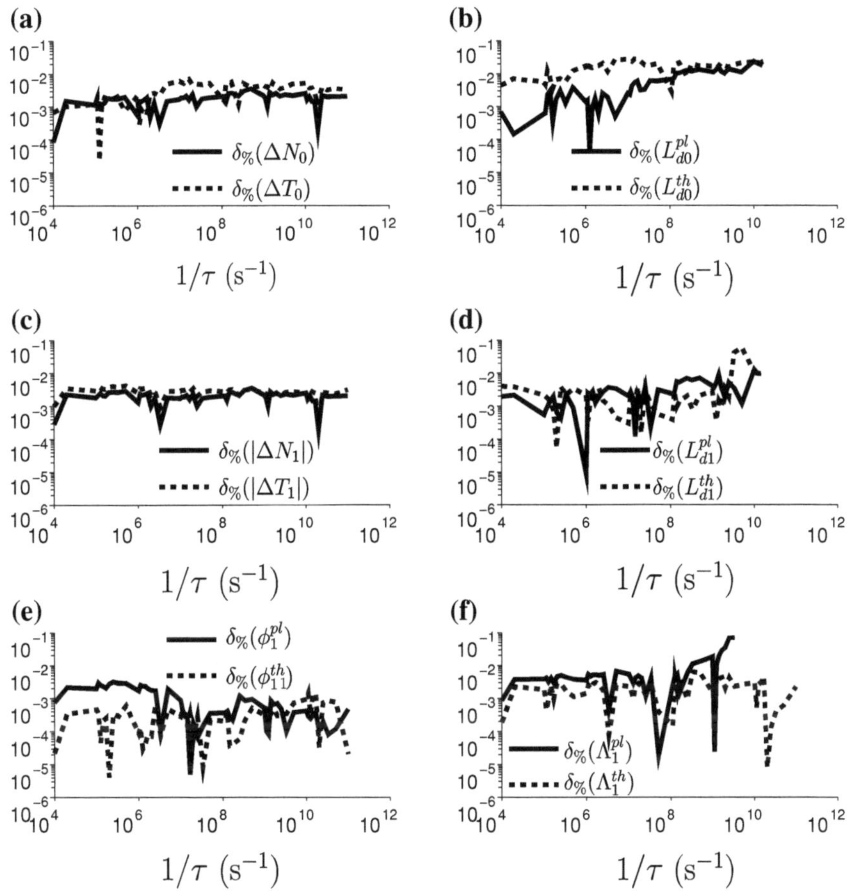

Fig. 4.13 Relative difference between the analytical and numerical solutions **a** of the *three-dimensional* ΔN_0 (*full line*) and ΔT_0 (*dashed line*) at position $(r, z) = (0, 0)$, **b** of their respective lateral decay lengths, **c** of the *three-dimensional* $|\Delta N_1|$ (*full line*) and $|\Delta T_1|$ (*dashed line*) at position $(r, z) = (0, 0)$, **d** of their respective lateral decay lengths, **e** of their respective phases and **f** of their respective lateral wavelengths in the case of the TP pump laser

is qualitatively very similar to its linear counterpart, it is still obvious that, from a quantitative viewpoint, it is significantly influenced by all the nonlinear effects.

Independently from the level of refinement, three distinct regimes can be observed. First, for low doping concentrations, a diffusion-limited regime can be observed. This is very interesting since it means that TP is only weakly sensitive to recombinations in a lowly doped Si substrate. It is therefore independent from the doping concentration in this region. Second, for intermediate doping concentrations ($>10^{17}$ cm^{-3}), the recombinations start to impact the carrier distributions. The unexpected bumpy behavior, not observed in the linear solutions, can be explained by a fine balance between the increase in recombinations with doping and the simultaneous decrease

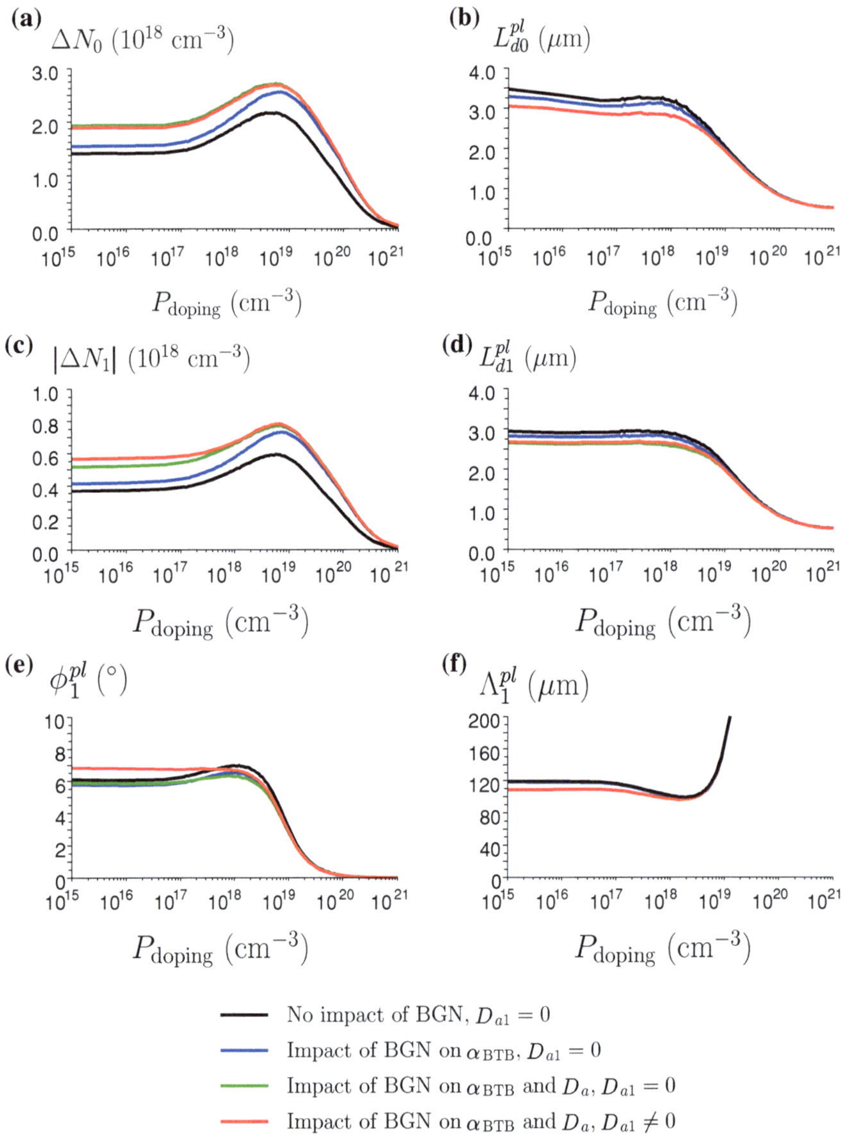

Fig. 4.14 Variation with p-type doping concentration **a** in the *three-dimensional* nonlinear ΔN_0 at position $(r, z) = (0, 0)$, **b** in its lateral decay length, **c** in the *three-dimensional* nonlinear $|\Delta N_1|$ at position $(r, z) = (0, 0)$, **d** in its lateral decay length, **e** in its phase and **f** in its lateral wavelength as obtained from the numerical resolution of Eqs. (4.54a–4.54d). Four levels of refinement of the model are shown. The *black lines* represent the solutions when BGN is neglected and only the constant mode of the ambipolar diffusivity is considered ($D_{a1} = 0$). The *blue* and *green lines* are obtained by adding the impact of BGN respectively upon the α_{BTB} of both lasers. The *green lines* are obtained by considering the impact of BGN upon D_a. Finally, the *red lines* show the solutions to the full equations, where the coupling due to D_{a1} is also considered

in ambipolar diffusivity. Finally, at high doping, a recombination-limited regime is reached.

Figure 4.15 shows the subsequent variations in ΔT_j ($j = 0, 1$) with doping. It can be seen that the temperature distributions are much less sensitive to the level of modeling refinement. The blue, green and red curves are superimposed, emphasizing that only the BGN-induced variations in absorption coefficient impact ΔT_j ($j = 0, 1$). Contrary to the excess carriers, the recombinations start to impact the temperature distribution for a relatively high doping concentration ($>5 \times 10^{18}\,\mathrm{cm}^{-3}$).

Before concluding this Section, we would like to discuss the dependence of the excess carrier distribution upon doping type. As it turns out, the temperature distribution indeed shows no sensitivity to doping type.

The variations in ΔN_j ($j = 0, 1$) with p-type (full lines) and n-type doping (interrupted lines) for our highest level of model refinement are shown in Fig. 4.16. Most important of all for the rest of this work, there is no sensitivity to doping type at low doping concentration (diffusion-limited regime).

When doping increases, however, the constant mode starts to distinguish between n- and p-type. Interestingly, for medium doping concentration, n-type doped Si shows a higher injection than p-type, while it is the opposite at high doping concentration. The difference at high doping concentration (recombination-limited regime) is readily explained by the larger Auger coefficient for n-type than p-type Si ($C_n \approx 3C_p$). The difference in behavior at moderate doping concentration is, however, more complex. Indeed, in the diffusion-limited regime found at low and moderate doping concentrations, it may seem surprising to observe that the excess carrier distributions on an n-type substrate have simultaneously higher amplitude and longer decay length. This effect is a direct consequence of the variation in ambipolar diffusivity shown in Fig. 4.5. These variations are indeed very different whether the sample is p- or n-type. First, the different excess carrier concentrations are easily explained by the different *average* ambipolar diffusivity. On average over the sample, the ambipolar diffusivity is indeed much larger on a p-type silicon sample, hence reducing the injection under the beam. Second, the different decay lengths are a direct consequence of the very fast variations in ambipolar diffusivity in the regions of medium injection in p-type silicon. These effects are therefore direct consequences of the strong nonlinearity of the ambipolar diffusivity in p-type silicon.

4.3 Summary

In this chapter, based on the thermodynamic model, we have derived simplified equations to explain quantitatively the behavior of the excess carriers and heat generated by the pump and probe lasers during a TP experiment on homogeneously doped silicon. We have shown that the resolution of a nonlinear system of four coupled equations (Eqs. (4.54a–4.54d)) is sufficient to determine the time- and spatial-variations in the excess carrier and temperature distributions. These four equations solve for the

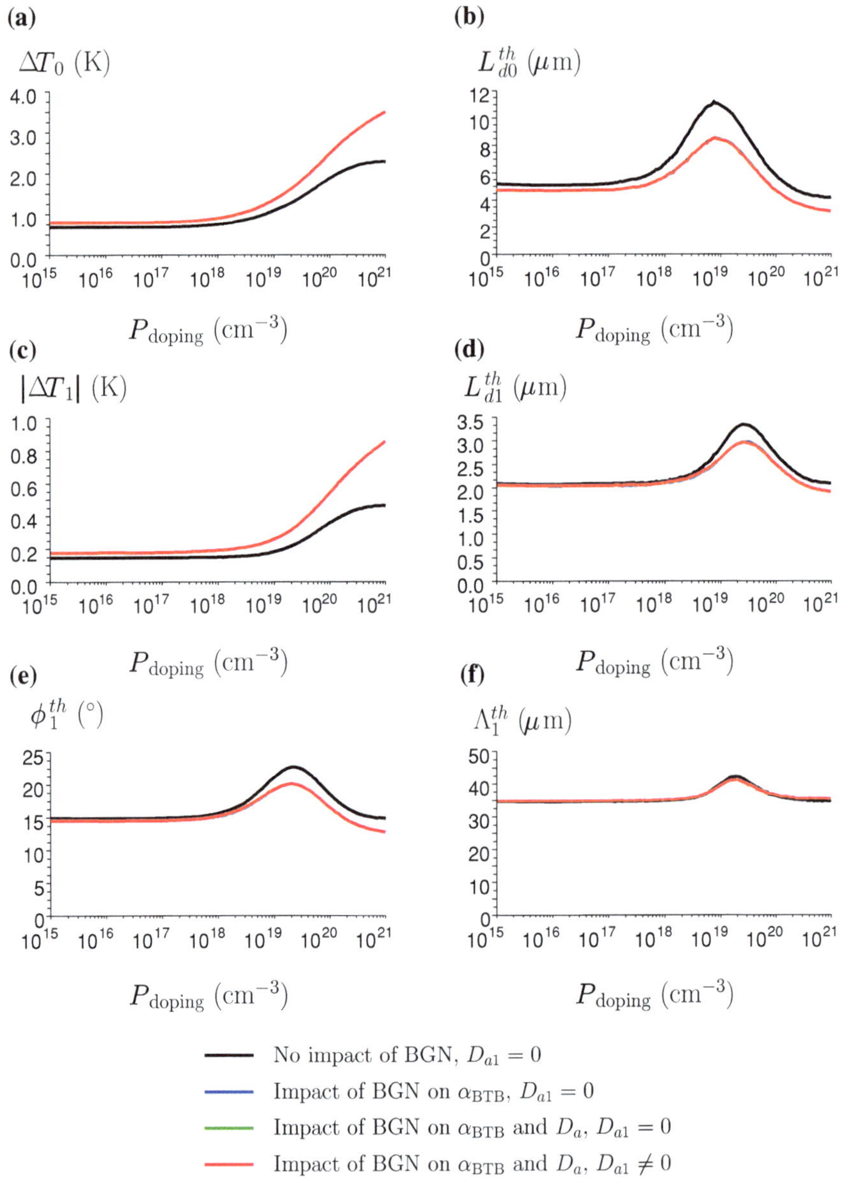

Fig. 4.15 Variation with p-type doping concentration **a** in the *three-dimensional* nonlinear ΔT_0 at position $(r, z) = (0, 0)$, **b** in its lateral decay length, **c** in the *three-dimensional* nonlinear $|\Delta T_1|$ at position $(r, z) = (0, 0)$, **d** in its lateral decay length, **e** in its phase and **f** in its lateral wavelength as obtained from the numerical resolution of Eqs. (4.54a–4.54d). Four levels of refinement of the model are shown. The *black lines* represent the solutions when BGN is neglected and only the constant mode of the ambipolar diffusivity is considered ($D_{a1} = 0$). The *blue* and *green lines* are obtained by adding the impact of BGN respectively upon the α_{BTB} of both lasers. The *green lines* are obtained by considering the impact of BGN upon D_a. Finally, the *red lines* show the solutions to the full equations, where the coupling due to D_{a1} is also considered

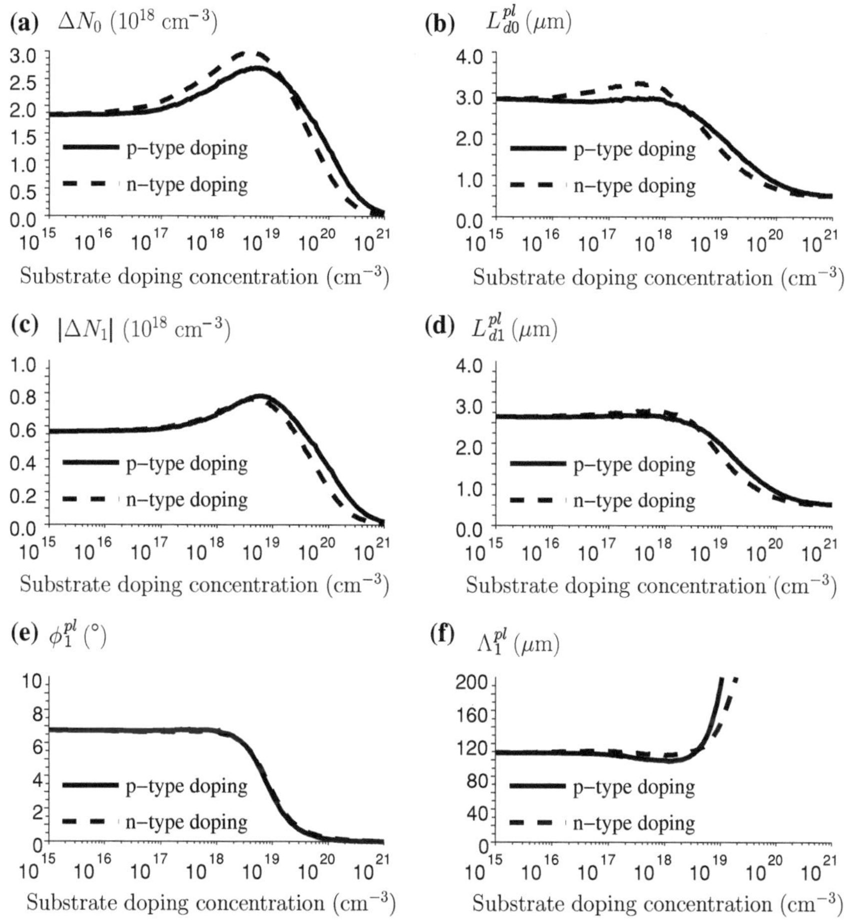

Fig. 4.16 Variation with p-type (*full lines*) and n-type (*dashed lines*) doping concentrations in **a** in the *three-dimensional* nonlinear ΔN_0 at position $(r, z) = (0, 0)$, **b** in its lateral decay length, **c** in the *three-dimensional* nonlinear $|\Delta N_1|$ at position $(r, z) = (0, 0)$, **d** in its lateral decay length, **e** in its phase and **f** in its lateral wavelength as obtained from the numerical resolution of Eqs. (4.54a–4.54d). Only the highest level of modeling refinement is shown, considering the impact of BGN upon the α_{BTB} of both lasers and upon D_a as well as the coupling due to D_{a1}

constant and fundamental modes of respectively the excess carrier and temperature distributions.

We have solved these equations and shown that the excess carriers and temperature behave typically as dampes waves, i.e. they are fully characterized by their amplitude, phase, decay length and wavelength (Eqs. (4.58a, 4.58b) and (4.59a, 4.59b)). Taking all nonlinear phenomena into account, the variations in these four parameters with the doping concentration of the silicon sample are shown in Fig. 4.14 for the excess carrier distribution and in Fig. 4.15 for the temperature distribution. The values obtained at low doping are of the highest importance for the rest of this work.

References

1. R.E. Wagner, A. Mandelis, Nonlinear photothermal modulated optical reflectance and photocurrent phenomena in crystalline semiconductors. 1. theoretical. Semicond. Sci. Technol. **11**(3), 289–299 (1996)
2. J. Opsal, Thermal and plasma waves in semiconductors. Rev Prog. Quant. Nondestruct. Evaluation **6B**, 1339–1346 (1987)
3. K. Kells, *General Electrothermal Semiconductor Device, Simulation* (Springer, Germany, 1994)
4. S.K. Sundaram, E. Mazur, Inducing and probing non-thermal transitions in semiconductors using femtosecond laser pulses. Nat. Mater. **1**(4), 217–24 (2002)
5. S. Selberherr, *Analysis and Simulation of Semiconductor Devices* (Springer, Wien, 1984)
6. A.H. Marshak, C.M. Vanvliet, Electrical-current and carrier density in degenerate materials with nonuniform band-structure. Proc. IEEE **72**(2), 148–164 (1984). ISSN 0018–9219
7. F. Dortu, J. Bogdanowicz, T. Clarysse, W. Vandervorst, Impact of band gap narrowing and surface recombination on photoelectrothermal modulated optical reflectance power curves. J. Vac. Sci. Technol. B **26**, 322–332 (2008)
8. S.M. Sze, *Physics of Semiconductor Devices*, 3rd edn. (Wiley-Interscience, New York, 2006)
9. J. McKelvey, *Solid-State and Semiconductor Physics*, Chap. 10 (Harper & Row, New York, 1966)
10. H.M. Vandriel, J.F. Young, Many-body effects and the diffusion of high-density electrons and holes in semiconductors. J. Phys. C **15**(3), L31–L35 (1982)
11. J.F. Young, H.M. Vandriel, Ambipolar diffusion of high-density electrons and holes in ge, si, and gaas—many-body effects. Phys. Rev. B **26**(4), 2145–2158 (1982). ISSN 0163–1829
12. A. Schenk, Finite-temperature full random-phase approximation model of band gap narrowing for silicon device simulation. J. Appl. Phys. **84**(7), 3684–3695 (1998)
13. F. Dortu, Low-Frequency Modulated Optical Reflectance for the One-Dimensional Characterization of Ultra-Shallow Junctions. PhD thesis, Katholieke Universiteit Leuven, 2009
14. F. Dortu, J. Bogdanowicz, Fsem, a semiconductor drift-diffusion equations solver using the finite elements method (fem), GNU GPL, (2005), http://fsem.sourceforge.net
15. F. Dortu, T. Clarysse, R. Loo, B. Pawlak, R. Delhougne, W. Vandervorst, Progress in the physical modeling of carrier illumination. J. Vac. Sci. Technol. B **24**(3), 1131–1138 (2006)
16. A. Mandelis, M. Nestoros, C. Christofides, Thermoelectronic-wave coupling in laser photothermal theory of semiconductors at elevated temperatures. Opt. Eng. **36**(2), 459–68 (1997)
17. A. Mandelis, *Diffusion-Wave Fields: Mathematical Methods and Green Functions*, 1st edn. (Springer, New York, 2001)
18. R.A. Smith, *Wave Mechanics of Crystalline Solids* (Chapman and Hall, London, 1969)
19. T. Yasuda, D.E. Aspnes, Optical-standard surfaces of single-crystal silicon for calibrating ellipsometers and reflectometers. Appl. Opt. **33**(31), 7435–7438 (1994)
20. D.E. Aspnes, A.A. Studna, E. Kinsbron, Dielectric-properties of heavily doped crystalline and amorphous-silicon from 1.5 to 6.0 ev. Phys. Rev. B **29**(2), 768–779 (1984)
21. D.B.M. Klaassen, A unified mobility model for device simulation. 1. model-equations and concentration-dependence. Solid-State Electron. **35**(7), 953–959 (1992)
22. J. Linnros, V. Grivickas, Carrier-diffusion measurements in silicon with a fourier-transient-grating method. Phys. Rev. B **50**(23), 16943–16955 (1994)
23. Y.P. Varshni, Band-to-band radiative recombination in groups iv, vi and iii-v semiconductors (i). Phys. Stat. Solidi **19**, 459–514 (1967)

24. M.E. Brinson, W. Dunstan, Thermal conductivity and thermoelectric power of heavily doped n-type silicon. J. Phys. C **3**, 483–491 (1970)
25. J. Bogdanowicz, F. Dortu, T. Clarysse, W. Vandervorst, D. Shaughnessy, A. Salnik, L. Nicolaides, J. Opsal, Advances in optical carrier profiling through high-frequency modulated optical reflectance. in Proceedings of The Insight Workshop , 2007)
26. H.S. Carslaw, J.C. Jaeger, *Conduction of Heat in Solids*, 2nd edn. (Oxford University Press, Oxford, 1986)
27. R.E. Collins, *Mathematical Methods for Physicists and Engineers* (Reinhold Book Corp., New York, 1968)
28. R.N. Bracewell, *The Fourier Transform and its Applications* (McGraw-Hill, New York, 1986)
29. M. Abramowitz, I. Stegun, *Handbook of mathematical functions* (Dover, New York, 1964)
30. M. Heath, *Scientific Computing, an Introductory Survey* (McGraw-Hill, New York, 1997)
31. J. Bogdanowicz, F. Dortu, T. Clarysse, W. Vandervorst, A. Salnik. Electrothermal theory of photomodulated optical reflectance on active doping profiles in silicon. J. Appl. Phys. **108**(10), 104908 (2010)
32. Synopsys. Medici, (2006), www.synopsys.com
33. Synopsys. Sentaurus device, (2006), www.synopsys.com

Chapter 5
Extension of the Transport Theory to Ultra-Shallow Doped Silicon Layers

In this chapter, we assume that a supra-bandgap modulated pump laser and a supra-bandgap constant probe laser shine on a *non*-homogeneously doped silicon sample with respective irradiances $\Pi_{\text{pump}}(x, y, t)$ and $\Pi_{\text{probe}}(x, y)$ such that

$$\begin{aligned}\Pi_{\text{pump}}(x, y, t) &= \mathcal{P}_{\text{pump}}(t) \exp\left(-\frac{x^2+y^2}{\mathcal{R}^2_{\text{pump}}}\right) \\ &= \mathcal{P}^0_{\text{pump}} \exp\left(-\frac{x^2+y^2}{\mathcal{R}^2_{\text{pump}}}\right)\left[\frac{1}{2}+\frac{1}{2}\cos(\omega t)\right] \\ &= \mathcal{P}^0_{\text{pump}} \exp\left(-\frac{x^2+y^2}{\mathcal{R}^2_{\text{pump}}}\right)\left[\frac{1}{2}+\frac{1}{4}\exp(i\omega t)+\frac{1}{4}\exp(-i\omega t)\right]\end{aligned} \tag{5.1}$$

$$\Pi_{\text{probe}}(x, y) = \mathcal{P}^0_{\text{probe}} \exp\left(-\frac{x^2+y^2}{\mathcal{R}^2_{\text{probe}}}\right), \tag{5.2}$$

where $\mathcal{P}_{\text{pump}}(t)$ is the time-dependent pump irradiance at position $(x, y) = (0, 0)$, $\mathcal{P}^0_{\text{pump}}$ (resp. $\mathcal{P}^0_{\text{probe}}$) is the peak value of the pump (resp. probe) irradiance, $\mathcal{R}_{\text{pump}}$ (resp. $\mathcal{R}_{\text{probe}}$) is the pump (resp. probe) radius and ω is the modulation frequency of the pump laser irradiance.

As a result, a free electron distribution $N(x, y, z, t)$, a free hole distribution $P(x, y, z, t)$ and a temperature distribution $T(x, y, z, t)$ are generated in the sample such that

$$\begin{aligned}N(x, y, z, t) &= N_{\text{doping}}(z) + \Delta N_0(x, y, z) \\ &+ \underbrace{\left[\Delta N_1(x, y, z)\exp(i\omega t) + \Delta N_1^\star(x, y, z)\exp(-i\omega t)\right]}_{\text{fundamental}}\end{aligned}$$

J. Bogdanowicz, *Photomodulated Optical Reflectance*, Springer Theses,
DOI: 10.1007/978-3-642-30108-7_5, © Springer-Verlag Berlin Heidelberg 2012

$$+\sum_{j=2}^{+\infty}\underbrace{\left[\Delta N_j(x,y,z)\exp(ij\omega t)+\Delta N_j^{\star}(x,y,z)\exp(-ij\omega t)\right]}_{j\text{th harmonic}} \tag{5.3}$$

$$\begin{aligned}P(x,y,z,t) = {} & P_{\text{doping}}(z)+\Delta P_0(x,y,z)\\ & +\underbrace{\left[\Delta P_1(x,y,z)\exp(i\omega t)+\Delta P_1^{\star}(x,y,z)\exp(-i\omega t)\right]}_{\text{fundamental}}\\ & +\sum_{j=2}^{+\infty}\underbrace{\left[\Delta P_j(x,y,z)\exp(ij\omega t)+\Delta P_j^{\star}(x,y,z)\exp(-ij\omega t)\right]}_{j\text{th harmonic}}\end{aligned} \tag{5.4}$$

$$\begin{aligned}T(x,y,z,t) = {} & T_0+\Delta T_0(x,y,z)\\ & +\underbrace{\left[\Delta T_1(x,y,z)\exp(i\omega t)+\Delta T_1^{\star}(x,y,z)\exp(-i\omega t)\right]}_{\text{fundamental}}\\ & +\sum_{j=2}^{+\infty}\underbrace{\left[\Delta T_j(x,y,z)\exp(ij\omega t)+\Delta T_j^{\star}(x,y,z)\exp(-ij\omega t)\right]}_{j\text{th harmonic}},\end{aligned} \tag{5.5}$$

where $N_{\text{doping}}(z)$ and $P_{\text{doping}}(z)$ are respectively the electron and hole concentrations due to the doping profile of the sample. Contrary to the previous chapter, these two carrier concentrations are here assumed to vary with depth. All the other quantities are as defined in Chap. 4.

This chapter extends the transport theory of carriers and heat discussed in Chap. 4 to the case of *actively doped silicon surfaces*. Throughout this chapter, we focus on the case of shallow doped layers on lowly doped silicon substrates, such as depicted in Fig. 5.1. In order to predict the behavior of the free carriers and heat generated in such a sample, one should in theory solve Eq. (4.6), with the consideration of the built-in electric field due to the uncompensated ionized dopant atoms in the space-charge region. The built-in electric field of *active* (i.e. annealed) doping profiles indeed modifies strongly the behavior of carriers in the region of the junction and in shallower regions. This obviously invalidates the ambipolar diffusion equation.

However, rather than solving directly Eq. (4.6), we propose to derive a simplified solution. Numerous advantages of this simplified solution can be given. First, obviously, the simplified solution makes the calculations much easier. This will prove very helpful, particularly for the inverse problem presented in Chap. 7. Second, this simplified solution also offers more understanding than a fully numerical solution. Third, the computation of a numerical solution to the considered problem would be very time consuming (time-dependent coupled nonlinear equations). Last but not

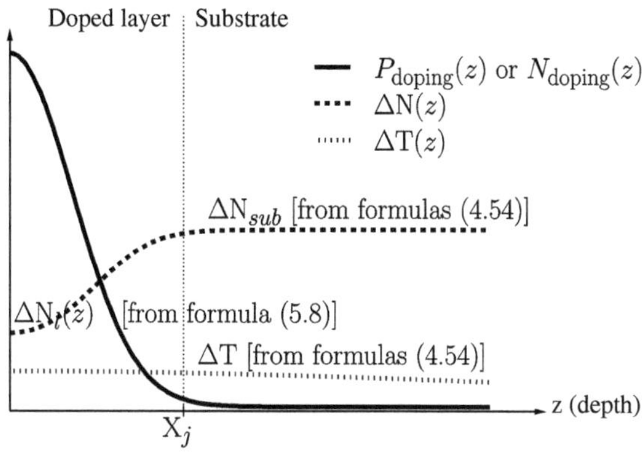

Fig. 5.1 Schematic of a typical active doping profile on a lowly doped substrate and the pump-generated excess carrier and excess temperature profiles (The relative position of the excess carrier and excess temperature profiles is arbitrary)

least, it also has to be pointed out that the numerical implementation of such a problem is very complex, as confirmed in the course of our investigations. Indeed, in spite of our attempts both on the full time-dependent and steady-periodic (Fourier domain) routes, we have not been able to force the system to converge. First, the time-dependent problem, where the time derivatives of Eq. (4.6) are discretized and the resolution is stepped over time, has proven to converge only for prohibitively small time steps, of the order of the dielectric relaxation time of silicon (1 ps), in agreement with [1]. The introduction of an artificial time-dependent term in Poisson's equation (4.6a), as suggested in Ref. [1], has been unfruitful. Second, in the steady-periodic system, where a separate equation is solved for each Fourier coefficient of the excess electron, hole and electrostatic potential, a problem also originates from the Poisson equation. The constant doping term $(N_d^+ - N_a^-)$ acts as a stabilizer for the steady equation. Such stability is, however, not present for the other Fourier coefficients, leading to a fast diverging system.

We here propose a third, simpler, approach. We show, in Sect. 5.1, that two assumptions suffice to strongly simplify the solutions to the carrier and heat transport problem on a sample with an ultra-shallow active doping profile (junction depth $X_j < 100$ nm), such as needed in modern CMOS transistors. In Sect. 5.2, we discuss quantitatively the validity of these assumptions on doping profiles with a box-like shape. The discussion is based on the comparison of the solutions obtained with steady drift-diffusion simulations (run with FSEM). In Sect. 5.3, we then derive the steady-periodic solutions under the same assumptions.

5.1 Simplified Transport Theory

As highlighted in the Introduction, we are seeking a simplified solution for the transport of carriers and heat in a sample presenting a doping profile such as shown in Fig. 5.1. For this purpose, we make two essential assumptions. First, we suppose that the electron and hole quasi-Fermi levels are flat through the space-charge region. Second, the carrier and temperature distributions in the bulk of the sample (substrate region of Fig. 5.1) are assumed independent from the doping profile. In this Section, we investigate qualitatively the validity of these assumptions and show how they simplify the problem.

First, we would like to study the flat quasi-Fermi levels approximation (FQL) and its implications.

Quasi-Fermi levels typically remain flat in highly conductive regions, whether the free carriers come from doping or optical injection. In order for the currents to flow [$|\vec{J_i}| = |\sigma_i \vec{\nabla} E^i_f|$ (Ref. [2]) where i = n or p], there is then indeed no need to strongly bend the quasi-Fermi levels. In other words, a semiconducting region with a high concentration of both electrons and holes usually shows flat quasi-Fermi levels.

The FQL approximation helps greatly when it comes to determining the excess carrier concentration in the doped layer. Indeed, the electrostatics, i.e. Poisson's equation (4.6a) or the p-n product [3], offers a direct link between the substrate and layer injection. We here use the p-n product which offers a simple solution. The p-n product reads, at each depth z of the sample,[1]

$$P(z)N(z) = \gamma_p(z)\gamma_n(z)n_i^2 \exp\left(-\frac{E_g(z)}{k_b T}\right) \exp\left(\frac{E_f^n(z) - E_f^p(z)}{k_b T}\right), \tag{5.6}$$

where γ_p and γ_n are the Fermi-Dirac factors respectively for holes and electrons defined in Eqs. (4.45a) and (4.45b). Using formula (5.6) both in the doped layer and in the substrate results in a simple expression for the excess carrier concentration in the layer. For this purpose, we assume here a p-type layer[2] with doping $P_{\text{doping}}(z)$. First, in the doped layer, the majority hole concentration is $P(z) = P_{\text{doping}}(z) + \Delta N_l(z)$ and the minority electron concentration is $N(z) \approx \Delta N_l(z)$. Second, in the highly injected substrate, independently from the doping type, $N(z) \approx P(z) \approx \Delta N_{sub}$. The depth-dependent excess carrier concentration can be obtained using the ratio of formula (5.6) taken respectively in the layer and in the substrate, i.e.

$$\frac{\left[P_{\text{doping}} + \Delta N_l(z)\right] \Delta N_l(z)}{\Delta N_{sub}^2} = \frac{\gamma_p^l(z)\gamma_n^l(z)}{\gamma_p^{sub}\gamma_n^{sub}} \exp\left(-\frac{E_g^l(z) - E_g^{sub}}{k_b T}\right), \tag{5.7}$$

where γ_n^l (resp. γ_p^l) and γ_n^{sub} (resp. γ_p^{sub}) are the electron (resp. hole) Fermi factor as defined above respectively in the layer and in the substrate, E_g^l and E_g^{sub} are respectively the bandgap energies in the layer and in the substrate. The only physically acceptable solution $\Delta N_l(z)$ of Eq. (5.7) is

---

[1] The dependence upon time is kept implicit so as to simplify the expressions.

[2] The case of an n-type layer can easily be derived by replacing P_{doping} by N_{doping} in the final expressions

$$\Delta N_l(z) = 0.5\left[-P_{\text{doping}}(z) + \sqrt{P_{\text{doping}}^2(z) + 4\frac{\gamma_p^l(z)\gamma_n^l(z)}{\gamma_p^{sub}\gamma_n^{sub}}\exp\left(-\frac{E_g^l(z) - E_g^{sub}}{k_b T}\right)\Delta N_{sub}^2}\,\right]. \quad (5.8)$$

From formula (5.8), one can show that $\Delta N_l(z)$ decreases with increasing $P_{\text{doping}}(z)$ in most cases. This explains the monotonically increasing excess carrier profile shown in Fig. 5.1.

Three implicit approximations have been made in the derivation of formula (5.8). First, the Debye length has been assumed much smaller than the characteristic lengths of the profile (junction depth and decay length). The used p-n product indeed supposes a vanishing Debye length, which is valid given the considered high doping concentrations and high injection. Second, the excess electron and hole concentrations have been assumed equal everywhere, in spite of the strong electric field at the junction. This is also related to the very short Debye length. The electric field is only present in a nanometer-wide region around the junction. In that region, the electrons and holes are in slightly different concentrations but this difference is invisible to long wavelengths such as that of the TP probe laser. Finally, we have also assumed that the substrate excess carrier concentration is flat (see Fig. 5.1) at the scale of X_j. This is related to the penetration depth of the pump and the plasma wave diffusion length being much longer than X_j (μm vs. nm).

In summary, the FQL approximation allows for a very attractive simplified derivation of the excess carrier concentration in the doped layer. As shown in formula (5.8), however, the substrate injection needs to be known.

The purpose of our second assumption, i.e. the independence of the substrate excess carrier concentration and temperature from the doped layer, is to give an easy access to the substrate injection. These two distributions are indeed assumed to be simply the solutions of the coupled ambipolar carrier diffusion and heat diffusion problem which has been solved in the previous chapter (e.g. Figs. 4.14, 4.15).

Actually, the presence of an actively doped layer at the surface of a silicon sample can modify the excess carrier and temperature in the bulk of the sample (i.e. below the doped layer) through three different phenomena. First, as a consequence of the presence of a doped layer, the generation rates of Eq. (4.52) increase both due to the slight reduction in the sample reflectance and to the rise in absorption coefficient (consequence of the combined Drude and BGN-induced electroabsorption effects as discussed in Sect. 3.2). However, as can be seen e.g. in Fig. 6.4a, the reflectance variations are very small (<1 %). The impact on the carrier and heat generation rates is therefore neglected. Similarly, the thicknesses of the considered ultra-shallow doped layer (<50–60 nm) being much smaller than the penetration depth of the probe and pump lasers (>5–15 μm), the increased absorption coefficient has hardly any impact on the final excess carrier and temperature distributions. Most of the absorption phenomenon indeed occurs in the bulk of the sample. Second, the presence of a doped layer at the surface, i.e. a zone of short recombination lifetime, acts as an enhanced surface recombination velocity. This effect tends to reduce the excess carrier concentration and increase the temperature. As we will see in Sect. 5.2.2, this effect is, however, never observed. This is also due to the deep penetration of the lasers compared to the thickness of the considered doped layers. Finally, the presence of a doped layer changes the electrostatic behavior of the substrate region close to the junction, leading to a possible small depletion of the sub-junction region. This last effect is dominant and actually the only one to ever be observed. We will show in Sect. 5.2.2 that the impact of this effect is limited on the carrier distributions and negligible on the temperature distribution.

Based on these two assumptions, the transport of optically generated free carriers and heat in USJs can be understood in the following way. First, concerning the heat, it is generated in the lowly doped substrate, where it diffuses according to Eqs. (4.54c) and (4.54d). Further, given that doping has a negligible impact on the thermal properties of Si at room temperature [4], the final excess temperature is layer-independent. Second, similarly, the excess carriers are generated in the substrate where they diffuse ambipolarly and recombine according to Eqs. (4.54a) and (4.54b). The final excess carrier distribution in the substrate is therefore layer-independent. The excess carrier concentration in the layer can be estimated by solving the one-dimensional in-depth Poisson's equation or, more simply, the p-n product assuming flat quasi-Fermi levels. In other words, the charging of the layer with excess carriers is taken care of solely by the electrostatics.

5.2 Validity of the Assumptions

Being convinced of the usefulness of our two simplifying assumptions, we would like to check quantitatively their validity. In order to do so, we propose to compare the simplified solutions with full drift-diffusion simulations, run with FSEM. More particularly, we base our discussion on 15-nm deep box-like profiles with varying doping concentrations. In Sect. 5.2.1, we look at the layer injection as a function of doping concentration in the layer and compare it with Eq. (5.8). We look at the impact of the presence of the doped layer on the substrate injection in Sect. 5.2.2.

Note that, as already mentioned in the Introduction of the present chapter, FSEM can only solve *steady* drift-diffusion equations (no waves). We therefore only consider this case. However, for the sake of consistency with the rest of this thesis, we would still like to separate the study into constant and fundamental modes of the excess carrier and temperature distributions. We use the superscript ^{DC} to stress that the distributions have been obtained using steady calculations. In other words, ΔN_0^{DC} and ΔT_0^{DC} have to be understood respectively as the excess carrier and temperature distributions obtained when the TP probe laser and the TP pump laser shine on the sample with respective *constant* irradiances $\mathcal{P}^0_{\mathrm{probe}}$ and $\mathcal{P}^0_{\mathrm{pump}}/2$. Similarly, ΔN_1^{DC} and ΔT_1^{DC} are the half-amplitudes of differential carrier concentration and temperature generated if the pump has an extra $\mathcal{P}^0_{\mathrm{pump}}/2$ *constant* irradiance.

Further, in order to clearly identify the different behavior of the carriers in the doped layer and in the substrate, we use the respective subscripts $_l$ and $_{sub}$ to refer to each region. This notation is actually of relevance in the rest of the thesis.

5.2.1 Flat Quasi-Fermi Level Approximation

In the case of a box-like profile, the excess carrier concentration ΔN_l in the doped layer is independent from depth and, following Eq. (5.8), it is linked to the excess carrier concentration ΔN_{sub} via

$$\Delta N_l = 0.5\left[-P_{\mathrm{doping}} + \sqrt{P_{\mathrm{doping}}^2 + 4\frac{\gamma_p^l \gamma_n^l}{\gamma_p^{sub}\gamma_n^{sub}}\exp\left(-\frac{E_g^l - E_g^{sub}}{k_b T}\right)\Delta N_{sub}^2}\,\right]. \tag{5.9}$$

This equation is valid at all lateral positions and can therefore also be used to determine the decay length.

The comparison of the excess carrier concentrations and decay lengths obtained respectively with FSEM simulations and the FQL approximation is shown in Fig. 5.2. The substrate injection ΔN_{sub}, needed in Eq. (5.9), has been obtained from the resolution of the steady ambipolar diffusion equation on a lowly doped substrate. The results are independent from the doping type of the substrate (not shown). It is clear from this figure that FQL is a very good approximation up to quite high doping concentrations ($>2 \times 10^{20}\,\mathrm{cm}^{-3}$). The values of the carrier concentrations and their decay lengths given by the FQL approximation indeed only depart from the numerical solution for doping concentrations higher than $2 - 3 \times 10^{20}\,\mathrm{cm}^{-3}$. Note that the excess carrier concentration in the layer then being approximately $10^{15}\,\mathrm{cm}^{-3}$, it is negligible. We therefore use FQL on the whole doping range, without any loss of accuracy in the model.

Two observations in Fig. 5.2a and c should be discussed. First, the different injection according to the doping type of the layer should be explained. It can be observed that, especially at high doping concentration, an n-type layer is more injected than a p-type layer with the same doping concentration. This difference is due to the combined effect of BGN and Fermi-Dirac statistics. Second, the discrepancy between the numerical solution and FQL at ultra-high layer doping concentrations should be looked at in more detail. As mentioned in the Introduction of this chapter, FQL is only valid in highly conductive regions. At high doping concentration, though the majority carrier conductivity is high, the minority carrier conductivity is low, leading to the possible bending of the quasi-Fermi level of the minority carriers. Note that, since an n-type doped layer is always more injected, FQL remains valid up to somewhat higher doping concentrations in such layers. It can be checked in Fig. 5.2

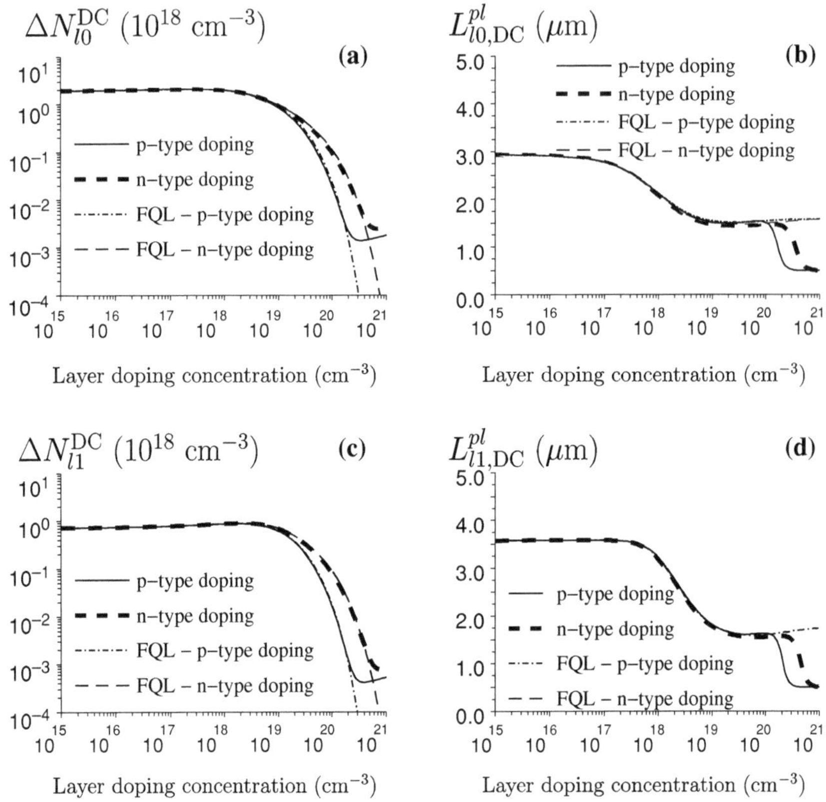

Fig. 5.2 Variation with n- and p-type layer doping **a** in $\Delta N_{l0}^{\mathrm{DC}}$ at position $(r, z) = (0, 0)$, **b** in its lateral decay length, **c** in $\Delta N_{l1}^{\mathrm{DC}}$ at position $(r, z) = (0, 0)$, **d** in its lateral decay length, as obtained by full three-dimensional steady drift-diffusion calculations (*full* and *thick interrupted lines*) and assuming flat quasi-Fermi levels (*thin interrupted lines*). Time-independence is assumed, as mentioned by the DC superscripts. The flat Fermi level approximation is in very good agreement with full calculations up to very high layer doping concentrations

that FQL becomes invalid at the same excess carrier concentration for p- and n-type layers ($\approx 10^{15}$ cm^{-3}).

Similarly, the decay lengths also show an interesting trend in Fig. 5.2b and d. As long as FQL is valid, the decay length follows a monotonic and slow decrease. This can be understood as a result of the lateral electric field felt by the excess carriers in the layer. It can indeed be shown that, independently from the doping type of the layer, the electric field always tends to push back the excess carriers towards the illuminated region, i.e. it goes against diffusion. Further, at the high doping concentrations invalidating FQL, the decay lengths drop very fast. This can be explained by the fact that electrostatics loses its influence on the layer injection. The excess carrier concentration in the layer therefore enters a recombination-limited regime (ultra-high doping), where the decay lengths are given by the pump radius.

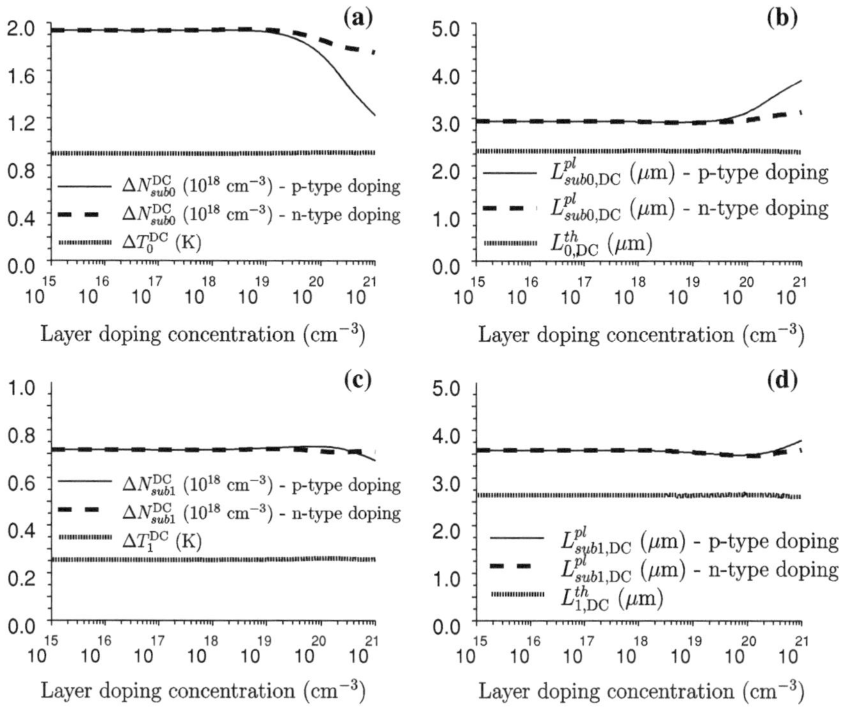

Fig. 5.3 Variation with n- (*interrupted lines*) and p-type (*full lines*) layer doping **a** in ΔN_{sub0}^{DC} and ΔT_0^{DC} (*dotted line*) at position $(r, z) = (0, 50)$ nm, **b** in their lateral decay lengths, **c** in ΔN_{sub1}^{DC} and ΔT_1^{DC} (*dotted line*) at position $(r, z) = (0, 50)$ nm, **d** in their lateral decay lengths, as obtained by full three-dimensional steady drift-diffusion calculations. Time-independence is assumed, as mentioned by the DC superscripts. The excess temperature does not depend upon the doped layer. The substrate injection is independent from the doped layer up to moderate doping concentrations

5.2.2 Impact of Doped Layers on Substrate Injection

In order to check quantitatively the variations in the excess carrier concentration and temperature in the substrate when a 15-nm thick doped layer is present at the surface, we propose to look at their behavior at a 50 nm depth from the surface, i.e. 35 nm deeper than the metallurgical junction. Figure 5.3 shows the variations with layer doping concentration in the excess carrier concentrations ΔN_{sub0}^{DC} and ΔN_{sub1}^{DC} but also in the excess temperature ΔT_0^{DC} and ΔT_1^{DC}, as obtained from three-dimensional steady drift-diffusion FSEM calculations.

It can be observed that, as expected, the doped layer has no impact whatsoever on the excess temperature. The excess carrier distribution, however, does vary in the high doping range. This effect is negligible on the fundamental mode ΔN_{sub1}^{DC} but not on the constant mode ΔN_{sub0}^{DC}, especially in the case of highly doped p-type layers (up to 20 % at a layer doping concentration of 2×10^{20} cm^{-3}). Note, however, that, as can be observed by the lower impact on $\Delta N^{\mathrm{DC}}_{sub1}$ than $\Delta N^{\mathrm{DC}}_{sub\,0}$, this effect depends to a great extent on the injection. Higher laser powers would advantageously reduce this effect significantly. Unfortunately, this effect cannot be accounted for by our model in its current status. We will come back to the impact of this approximation in Sect. 6.3.

The origin of this effect, as already hinted at, cannot be recombinations. Auger recombinations, i.e. the dominant recombination effect in highly doped layers, are indeed more numerous in n-type silicon than in p-type silicon [5] ($C_n \approx 3C_p$). Besides, as confirmed by Fig. 5.2, n-type layers are more injected than p-type layers, which should further increase the recombinations in n-type layers. The slight depletion of the substrate actually originates from the lateral junction photopotential due to the non-uniform illumination of the built-in electric field. This photopotential perturbs the behavior of the carriers in the substrate. The associated lateral electric field points into the outward radial direction for a p-type layer and inward radial direction for an n-type layer. The Dember field in the substrate, however, always points into the outward radial direction. The Dember field can therefore slightly counteract the lateral junction photopotential for an n-type but not a p-type layer.[3]

For the sake of completeness, let us add that the values shown in Fig. 5.3 are independent from the doping type of the substrate (not shown).

5.3 Steady-Periodic Model Equations

So far the discussion has been limited to steady calculations due to the limitations of FSEM, our numerical solver. However, the simplified theory can easily be extended to the steady-periodic calculations needed in this work. In this Section, we assume that the substrate injection is, independently from the layer doping, as calculated in Chap. 4 for a substrate doped homogeneously with a $10^{15}\,\mathrm{cm}^{-3}$ concentration. Based on this substrate injection, we investigate how the constant mode ΔN_{l0} and fundamental mode ΔN_{l1} of the excess carrier distribution[4] in a box-like doped layer vary as a function of the active doping concentration, if we assume flat quasi-Fermi levels through the space-charge region.

More specifically, we consider a substrate injection $\Delta N_{sub} = \Delta N_{sub0} + 2\,\Re[\Delta N_{sub1}\exp(i\omega t)]$ where $\Delta N_{sub\,0}$ and ΔN_{sub1} behave as found in the low doping region of Fig. 4.14 (complete model, i.e. red curves). As a consequence of the independence from substrate doping concentration of the latter two quantities in

[3] Canceling the Dember field by setting equal electron and hole mobilities results in equally lowered substrate injection for n- and p-type layers.

[4] Note that, as we have in Chap. 4, we should also investigate the generation of further harmonics. However, only the constant and fundamental modes of the layer excess carrier concentration contribute to the measured signals (see Chap. 6). The other harmonics are therefore neglected. The reason for the in-depth discussion of the harmonic generation in Chap. 4 is that the existence of further harmonics in the substrate carrier distribution would have modified the layer excess carrier concentration via Eq. (5.9)

the doping range below $5 \times 10^{17}\,\text{cm}^{-3}$, the obtained ΔN_{l0} and ΔN_{l1} are equally independent from doping in that range. The results are obviously also independent from the doping type of the substrate (Fig. 4.16). We then make use of Eq. (5.9) to determine the layer injection.

The time-dependence of the considered substrate injection does actually not make the situation more complex than the steady case studied thus far. Indeed, all the quantities present in Eq. (5.9) can accommodate the (slow) variations in substrate injection without any lag. They take at most 1 picosecond to follow the variations (dielectric relaxation and relaxation time of the Fermi-Dirac quasi-equilibrium distributions) [6]. In conclusion, it can be expected that the variations in excess carrier concentration in the layer are simultaneous with the excess carrier concentration in the substrate. This further assumes that the layer thickness is small compared to the plasma wavelength Λ_1^{pl}, which is obviously the case in this work (nm vs. hundreds of μm, see Fig. 4.14).

Figure 5.4 shows the calculated variations in ΔN_{l0} and ΔN_{l1} as a function of p- and n-type doping concentrations in the layer. The behavior of the steady distributions of Fig. 5.2 is clearly recognized and their discussion therefore still applies here. Further, the phase and wavelength shown respectively in Fig. 5.4e and f prove to be independent from the layer doping concentration and type. They are actually the same as in the substrate, as supposed above. In other words, the fundamental mode of the excess carrier distribution behaves as a single wave, just as it does in a homogeneously doped substrate. The only peculiarity of the wave when a doped layer is present is that its amplitude and lateral decay length strongly and abruptly decrease in the highly doped regions. The subscript $_l$ and $_{sub}$ for the phase and wavelength of the plasma wave are therefore dropped in the rest of this thesis, since a single value suffices.

The results shown in Fig. 5.4 combined with the results of Figs. 4.14 and 4.15 summarize with great simplicity our model of the carrier and heat transport in non-homogeneously doped silicon.

Before concluding, we believe it is important to derive the following approximate expression for ΔN_{l1} at high p-type[5] doping concentrations ($>10^{19}\,\text{cm}^{-3}$), asssuming Boltzmann statistics and a time-independent bandgap. Under these approximations, based on Eq. (5.9), it can be shown that

$$\Delta N_{l1} \approx \frac{2\Delta N_{sub\,0}\Delta N_{sub\,1}}{P_{\text{doping}}} \exp\left(-\frac{E_g^l - E_g^{sub}}{k_b T}\right). \tag{5.10}$$

This expression is of critical importance for the general understanding of the behavior of ΔN_{l1}. It shows in very simple mathematical terms e.g. that the fundamental mode of the excess carrier distribution in the layer drops with the active doping concentration in this layer. We will come back to this observation in Sect. 6.2 of the next chapter. Furthermore, it also demonstrates the dependence of ΔN_{l1} upon

[5] This equation is easily extended to n-type doped layers by replacing P_{doping} by N_{doping}

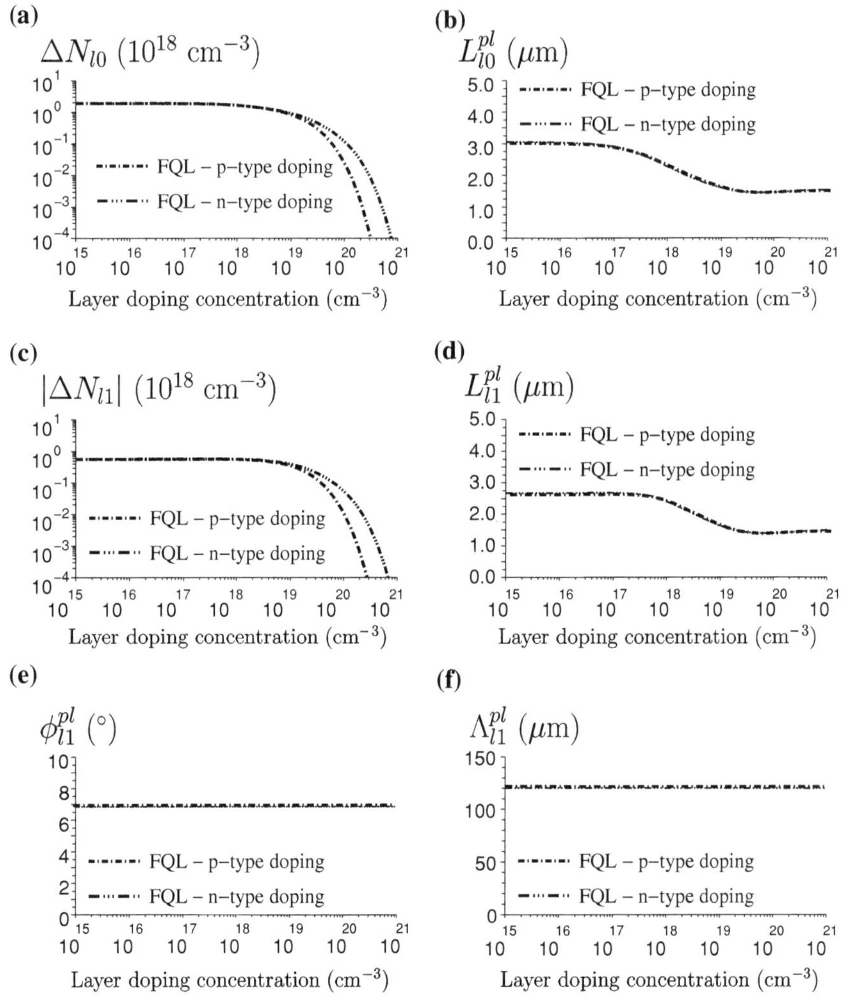

Fig. 5.4 Variation with n-type (*triple interrupted lines*) and p-type (*interrupted lines*) doping concentration **a** in ΔN_{l0} at position $(r, z) = (0, 0)$, **b** in its lateral decay length, **c** in $|\Delta N_{l1}|$ at position $(r, z) = (0, 0)$, **d** in its lateral decay length, **e** in its phase and **f** in its lateral wavelength as obtained assuming flat quasi-Fermi levels and the substrate carrier concentrations found in the low doping region of Fig. 4.14

ΔN_{sub0}. In other words, though time-independent, ΔN_{sub0} is also involved in the fundamental mode of the excess carrier distribution via ΔN_{l1}. This comment will have its importance for the understanding of the modeling error discussed in Sect. 6.3 of the next chapter.

5.4 Summary

In this chapter, we have extended the transport theory developed in the previous chapter to samples presenting a shallow doped layer at their surface.

First, we have shown that the temperature distribution is independent from the doped layer. Second, we have shown that the fundamental mode of the excess carrier distribution below the doped layer (substrate region) is also independent from the layer. The constant mode, however, has shown non-negligible variations at high layer doping concentrations. The impact of this approximation is looked at in Sect. 6.3. Furthermore, we have demonstrated that, if the quasi-Fermi levels of the electrons and holes remain flat in the structure, the excess carrier distribution in the doped layer is readily derived from the excess carrier distribution in the substrate via Eq. (5.8). Most importantly, Fig. 5.4 shows the behavior of ΔN_{l0} and ΔN_{l1} as a function of the n-type and p-type doping concentrations in the layer assuming low doping in the substrate and the excess carrier concentration derived in Chap. 4 (Fig. 4.14).

References

1. M.S. Mock, *Analysis of Mathematical Models of Semiconductor Devices* (Boole Press, Dublin, 1983)
2. A.H. Marshak, C.M. Vanvliet, Electrical-current and carrier density in degenerate materials with nonuniform band-structure. Proc. IEEE **72**(2), 148–164 (1984). ISSN 0018–9219
3. A.H. Marshak, Transport-equations for highly doped devices and heterostructures. Solid-State Electron. **30**(11), 1089–1093 (1987)
4. R. Hull (ed.) *Properties of Crystalline Silicon* (INSPEC, London, 1999)
5. S. Selberherr, *Analysis and Simulation of Semiconductor Devices* (Springer, Wien, 1984)
6. S.K. Sundaram, E. Mazur, Inducing and probing non-thermal transitions in semiconductors using femtosecond laser pulses. Nat. Mater. **1**(4), 217–24 (2002)

Chapter 6
Assessment of the Model

In the previous chapters, we have developed a model to explain quantitatively how the reflectance of a silicon sample can be modified by doping and/or laser-injected carriers and heat. Going backwards from Chaps. 5 to 2, we have sequentially calculated the excess carrier and temperature distributions generated by the TP pump and probe lasers in Chaps. 4 and 5 respectively for homogeneously and non-homogeneously doped silicon. The perturbation of the refractive index induced by these distributions as well as doping have been derived in Chap. 3. Finally, we have calculated the relationship between the perturbation of the refractive index profile and the perturbation of the sample reflectance in Chap. 2. Though this theoretical investigation has offered much insight, the quality of this model still needs to be assessed. In this chapter, we therefore evaluate the accuracy of the model developed in the previous chapters on two types of samples. In Sect. 6.1, we sum up and test the model for homogeneously doped silicon samples as a function of doping concentration. In Sect. 6.2, we then study the case of shallow doped layers. For the sake of simplicity, we only investigate profiles with a simple and perfectly defined shape, i.e. box-like doping profiles, as can be obtained from Chemical Vapor Deposition[1] (CVD). For these samples, we also summarize the model and check its validity as a function of the thickness and doping of the layer. Note that only lowly doped substrates are considered (i.e. no well, no halo). Finally, in Sect. 6.3, we identify the errors made by our model and their most likely sources.

In this work, we measure the so-called DC reflectance R_{dc} and the so-called modulated—or AC—reflectance ΔR_{ac}. More particularly, we measure their offset curves, i.e. we measure these two signals as a function of the spacing between the probe and the pump beams, denoted r in the present chapter. The DC reflectance is the total time-independent reflectance, i.e.

$$R_{\mathrm{dc}}(r) = R_0 + \Delta R_0(r), \tag{6.1}$$

[1] I would like to thank Duy Nguyen and Laurent Souriau for the layers they have fabricated for this study.

DOI: 10.1007/978-3-642-30108-7_6,

where ΔR_0 includes all time-independent contributions to the reflectance perturbation, i.e. doping, steady excess carrier and steady temperature effects. As for ΔR_{ac}, it is the normalized fundamental mode of the reflectance perturbation,[2] i.e.

$$\Delta R_{ac}(r) = \Gamma_0 \exp(-i\theta_0)\Delta R_1(r), \tag{6.2}$$

where ΔR_1 includes the time-dependent contributions to the reflectance perturbation, i.e. the fundamental mode of the excess carrier and temperature distributions. Γ_0 and θ_0 are respectively the normalization coefficient and phase of the fundamental mode of the reflectance perturbation. These two factors originate from the lock-in amplifier and are unfortunately unknown.

As we will show, the understanding of the behavior of R_{dc} is quite straightforward. On the contrary, ΔR_{ac} offset curves are complex (i) to measure with a sufficient repeatability, (ii) to understand and (iii) to model. The three remarks below are therefore of critical importance for the understanding of this chapter.

(i) As studied in Refs [1, 2], ΔR_{ac} varies with illumination time when measured on crystalline Si. As a consequence, the measurement of ΔR_{ac} suffers from a very poor repeatability if no attention is paid to this issue. A procedure ensuring the high measurement repeatability of ΔR_{ac} has therefore been developed and systematically used in the framework of our investigations. As detailed in Appendix A.1, in this procedure, each offset curve is composed of 25 data points, each obtained after shining the pump and probe beams for 25 s on the sample. Each offset curve is measured initially with aligned beams (0 separation) and by separating the beams of 0.167 μm every 25 s. This allows us to compromise between measurement repeatability [$\sigma_{\Delta R_{ac}} = 7 \times 10^{-4}$ (a.u.), Appendix A.2] and measurement time (≈10 min/curve[3]). Further, all our CVD boxlike profiles have been grown on device-quality wafers. This ensures a sufficient wafer-to-wafer repeatability of the signals. This complete procedure is of critical importance since the good agreement which our theory will show with experimental ΔR_{ac} in this chapter would not be possible without it.

(ii) As can already be expected from Eq. (6.2), ΔR_{ac} is the superposition of multiple damped waves and therefore has a complex behavior. To facilitate the comparison between experimental and theoretical data, it is convenient to also write it as a damped wave, i.e.

$$\begin{aligned}\Delta R_{ac}(r) &= |\Delta R_{ac}(r)| \exp\left[-i\phi_{ac}(r)\right] \\ &= |\Delta R_{ac}(r=0)| \exp\left[-i\phi_{ac}(r=0)\right] \exp(-r/L_d^{ac}) \exp(-2i\pi r/\Lambda_{ac}),\end{aligned} \tag{6.3}$$

[2] Note that, contrary to Eq. (2.2), ΔR_1 is here defined in its complex number notation. We are aware of this slight inconsistency. It is, however, more convenient to write ΔR_1 and, consequently, ΔR_{ac}, in complex notation so that the link between the behavior of the signal and its components is more readily observed.

[3] I want to thank Derrick Shaughnessy again for writing a measurement routine for the TP tool thanks to which an uncountable number of hours have been saved.

where $|\Delta R_{ac}(r = 0)|$ and $\phi_{ac}(r = 0)$ are respectively the measured amplitude and phase of the AC reflectance under the pump beam, while L_d^{ac} and Λ_{ac} are respectively the signal decay length and wavelength. By definition, L_d^{ac} is the lateral distance needed for the amplitude $|\Delta R_{ac}|$ to drop by a factor exp(1), i.e. it is linked to the slope of the offset curve of the signal *amplitude*. We calculate it from the following equation

$$L_d^{ac} = \left. \frac{-|\Delta R_{ac}|}{\frac{\partial |\Delta R_{ac}|}{\partial r}} \right|_{r=0}, \tag{6.4}$$

which can be deduced from Eq. (6.3). As for the signal wavelength Λ_{ac}, it is the lateral distance needed for the phase ϕ_{ac} to turn 360°, i.e. it is linked to the slope of the offset curve of the signal *phase*. We calculate it by extrapolation of the offset curves, i.e.

$$\Lambda_{ac} = \frac{360}{\phi_{ac}(r = 4\,\mu\text{m}) - \phi_{ac}(r = 0)} \times 4(\mu\text{m}). \tag{6.5}$$

The term 'lengths' used for L_d^{ac} and Λ_{ac} is somewhat abusive since negative values of both the signal decay length and the signal wavelength are allowed. This a priori unphysical possibility is just a consequence of their mathematical definitions. Obviously, the respective decay lengths and wavelengths of each damped wave contributing to the signal are all positive, i.e. their respective amplitudes decrease and their respective phases increase. A backward diffusion wave going against the diffusion force is indeed not physical [3]. But the AC reflectance is a combination of these damped waves. Subsequently, its amplitude $|\Delta R_{ac}|$ can, in some specific occasions, rise with laser beam separation,[4] i.e. the signal decay length can be negative [4]. Similarly, the phase ϕ_{ac} can sometimes decrease with laser beam separation, i.e. the signal wavelength can be negative. We show below that, albeit unexpected, these situations are observed experimentally and can be explained with the introduced model.
Note finally that, rigorously, ΔR_{ac} is the sum of multiple damped waves. Equation (6.3) is therefore only an approximation. This approximation quantitatively represents the measured behavior of ΔR_{ac} offset curves in most cases. As we will see, however, it fails to do so in some complex cases (moderately doped CVD layers). In spite of this limitation, these equations help significantly to understand the complex behavior of ΔR_{ac} offset curves, which justifies their use in this work.

(iii) As a direct consequence of the separated pump and probe beams during the measurement of an offset curve, their positions of maximum irradiance no longer coincide, which slightly modifies the distributions of the generated excess car-

[4] Note that this can obviously only be true for a limited region of the offset curve, i.e. the amplitude does not diverge. What is observed in these cases is that the rise in amplitude is only sustained in the first few micrometers and is followed by the expected decaying amplitude.

riers and excess temperature. We, however, neglect these variations. In other words, our calculations consider that the probe and pump lasers always shine on the same spot, generating the carriers and temperature exactly as discussed in Chaps. 4 and 5. A third, fictitious, laser of wavelength λ_{probe} is assumed to measure the offset curve without generating any excess carrier or heat. Considering the separation between the probe and pump lasers in the numerical calculation of the excess carriers and heat would break the central symmetry of the problem and therefore require a full three-dimensional treatment. Unfortunately, FSEM does not have that capability. This approximation is very good, owing to the constancy and relatively low magnitude of the probe laser irradiance. Note also that the radius of this fictitious laser is assumed sufficiently small to measure the local (and not integrated) variations in reflectance (see Sect. 2.5.2). It can be checked that considering the finite dimension of the probe radius would yield slightly better results (mostly for L_d^{ac} and Λ_{ac}) but would not change the final conclusions of this chapter.

6.1 Homogeneous Doping

Before looking at experimental data measured on homogeneously doped silicon substrates, let us summarize the model for such samples. Two perturbation effects are to be taken into account. First, when the homogeneous doping concentration is increased, the refractive index is reduced, as studied in Chap. 3. In turn, Chap. 2 has shown that this reduces the sample reflectance. Second, when the pump and probe lasers shine on the homogeneously doped sample, excess carrier and excess temperature distributions such as those derived in Chap. 4 are generated (see, in particular, Figs. 4.14 and 4.15). These distributions also perturb the refractive index of the sample, thus modifying the sample reflectance. Since the excess carrier and temperature distributions have characteristic lengths longer than the probe wavelength, the perturbation of the reflectance is only due to the surface perturbation of the refractive index. In other words, using Eqs. (2.11) and (3.16), R_{dc} can be written

$$R_{\text{dc}}(r) = R_0 \left\{ 1 + \frac{4}{n_0^2 - 1} \left[-\beta \left(\frac{1}{m_h} P_{\text{doping}} + \left(\frac{1}{m_e} + \frac{1}{m_h} \right) \Delta N_0(r) \right) + \delta \Delta T_0(r) \right] \right\}, \tag{6.6}$$

for a homogeneous p-type doping concentration P_{doping}. The laser-induced variations in DC reflectance due to the time-independent excess carrier and temperature distributions are never observed (see e.g. Fig. 6.1a), due to the low magnitudes of the steady excess carriers and temperature $[(1/m_e + 1/m_h)\beta \Delta N_0 \ll 1$ and $\delta \Delta T_0 \ll 1]$. We can therefore use the following simplified expressions

$$R_{\text{dc}} = R_0 \left[1 - \frac{4\beta}{m_h(n_0^2 - 1)} P_{\text{doping}} \right] \quad \text{for p-type doping} \tag{6.7a}$$

$$R_{dc} = R_0 \left[1 - \frac{4\beta}{m_e(n_0^2 - 1)} N_{doping} \right] \text{ for n-type doping.} \tag{6.7b}$$

Similarly, using the same Eqs. (2.11) and (3.16), the AC reflectance can be written

$$\begin{aligned} \Delta R_{ac}(r) &= \frac{4R_0}{n_0^2 - 1} \Gamma_0 \exp(-i\theta_0) \left[-\beta \left(\frac{1}{m_e} + \frac{1}{m_h} \right) \Delta N_1(r) + \delta \Delta T_1(r) \right] \\ &= \frac{4R_0}{n_0^2 - 1} \Gamma_0 \exp(-i\theta_0) \\ &\times \Bigg[\underbrace{-\beta \left(\frac{1}{m_e} + \frac{1}{m_h} \right) |\Delta N_1(r=0)| \exp(-i\phi_1^{pl}(r=0)) \exp(-r/L_{d1}^{pl}) \exp(-2i\pi r/\Lambda_1^{pl})}_{\text{plasma component}} \\ &\quad \underbrace{+ \delta |\Delta T_1(r=0)| \exp(-i\phi_1^{th}(r=0)) \exp(-r/L_{d1}^{th}) \exp(-2i\pi r/\Lambda_1^{th})}_{\text{thermal component}} \Bigg]. \end{aligned} \tag{6.8}$$

Note that, though $|\Delta N_1|$ and $|\Delta T_1|$ have the same order of magnitude as ΔN_0, their modulation allows for the use of a lock-in amplifier. As a result, contrary to ΔN_0, their impact on the sample reflectance can be measured with a high reproducibility.

Equations (6.7) and (6.8) together with the excess carrier and temperature distributions respectively shown in Figs. 4.14 and 4.15 (complete model, i.e. red curves) constitute our model for homogeneously doped silicon. In summary, the variations in R_{dc} are only due to doping. As for ΔR_{ac}, it consists of two components respectively related to the fundamental modes of the surface excess carriers and temperature generated in the silicon substrate. We call these two components respectively the *plasma* and *thermal* components. Both components vary with doping as studied in Chap. 4. Very importantly for the understanding of the behavior of the total signal, these two components have opposite signs and different phases, i.e. they show a total phase difference close to 180° but not precisely equal to 180°.

6.1.1 Comparison of the Model with Experimental Data

Let us start by presenting the experimental offset curves and discuss qualitatively their behavior when the doping concentration varies in Sect. 6.1.1.1 below. The model is then compared quantitatively to the measurements in Sect. 6.1.1.2.

6.1.1.1 Qualitative Comparison

Figures 6.1a–c show experimental offset curves of respectively R_{dc}, $|\Delta R_{ac}|$ and ϕ_{ac} for five different p-type doping concentrations.

Starting with Fig. 6.1a, it is striking that R_{dc} does not vary with beam separation. This shows that ΔN_0 and ΔT_0 do not impact R_{dc} significantly, which

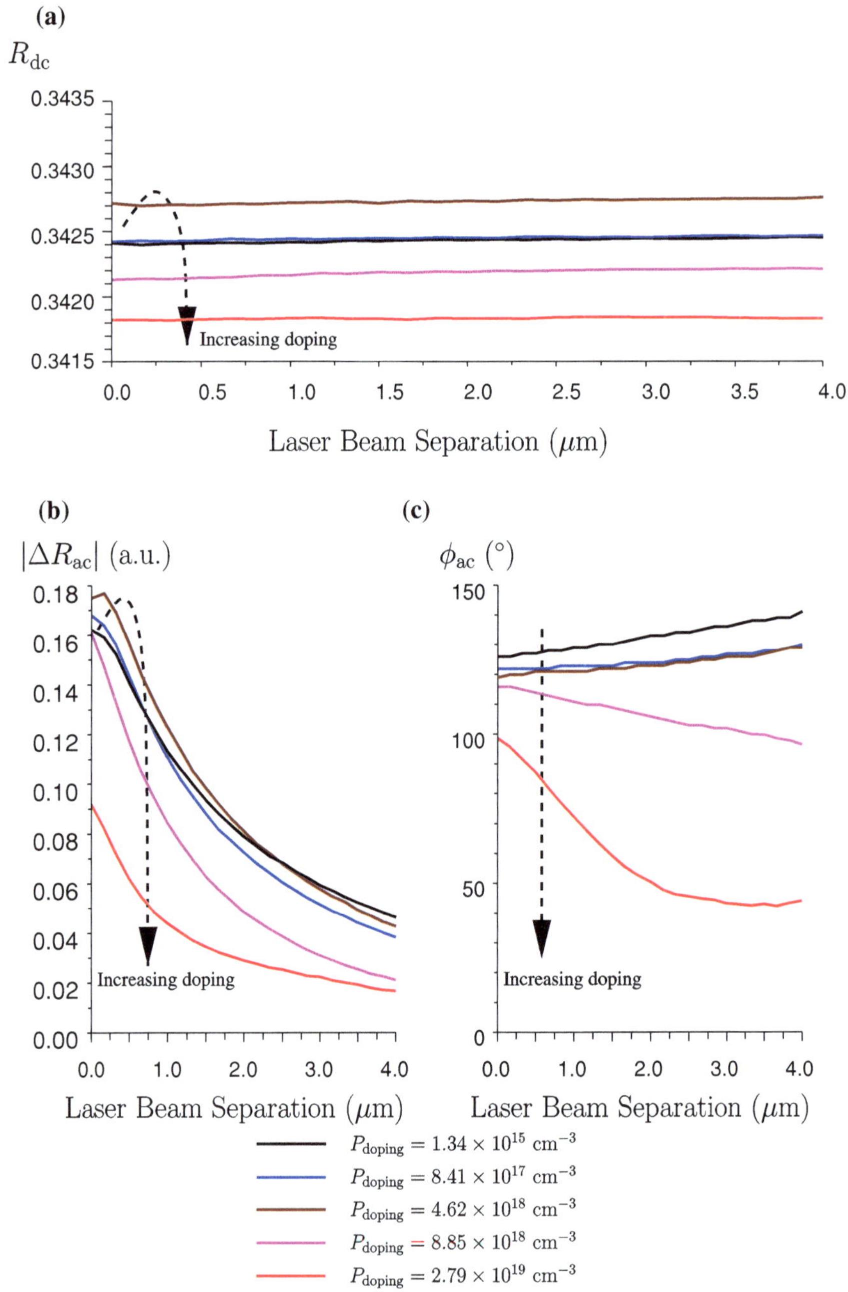

Fig. 6.1 Experimental offset *curves*, i.e. variation with laser beam separation, of **a** the DC reflectance R_{dc}, **b** the amplitude and **c** phase of the AC reflectance ΔR_{ac} for five homogeneously p-doped substrates with doping concentrations ranging from 10^{15} to $3 \times 10^{19}\,cm^{-3}$. The *arrows* show the trend observed when increasing doping

justifies our simplified model Eqs. (6.7a) and (6.7b). Further, the experimentally observed behavior of R_{dc} for increasing doping concentration follows the monotonic decrease expected by Eq. (6.7a). We, however, note one outlier (see higher R_{dc} for $P_{doping} = 4.62 \times 10^{18}\,\text{cm}^{-3}$). The existence of this outlier is not a problem in itself but it underlines how careful one should be when trying to understand or use a signal like R_{dc}. This signal indeed shows very small variations (<1 %) when doping changes and, as a consequence, it can be strongly impacted by any other effect. Unfortunately, all our homogeneously doped samples are old and have not been monitored over the years.[5] The reasons for the unexpectedly high R_{dc} on this particular outlier are therefore not clear. We have actually noted quite a few outliers in our database of homogeneously doped substrates. The study of the DC reflectance on homogeneously doped substrates therefore stops here.

The offset curves of $|\Delta R_{ac}|$ shown in Fig. 6.1b show the clear fingerprint of the plasma component of Eq. (6.8). Indeed, the non-monotonic behavior of $|\Delta R_{ac}(r = 0)|$ for coincident beams when doping increases is in excellent agreement with the behavior of $|\Delta N_1|$ in Fig. 4.14c. Interestingly, these experimental data provide evidence for the theoretically predicted bump of Fig. 4.14c in the $10^{18} - 10^{19}\,\text{cm}^{-3}$ doping range. Similarly, the experimental signal decay length is observed to decrease monotonically with doping, in good agreement with the behavior of L_{d1}^{pl} in Fig. 4.14d.

The discussion of Fig. 6.1c is slightly more complex since the impact of the thermal component of Eq. (6.8) is here more apparent. The phase measured for coincident beams $\phi_{ac}(r = 0)$ decreases monotonically, such as expected from Fig. 4.14e. However, the measured phase variation is much larger than that calculated for the plasma component alone (>30° vs. 7°). Actually, what is observed is the progressive transition from a plasma-dominated behavior at low doping concentration to a thermally dominated behavior at high doping concentration. Similarly, the signal wavelength increases with doping for low and medium doping ($\leq 4.62 \times 10^{18}\,\text{cm}^{-3}$) as expected from a purely plasma signal (Fig. 4.14f). However, at high doping it reaches negative values, which could not be explained without the consideration of the thermal component. The wavelength would then indeed always be positive. The negative signal wavelength at high doping is actually a consequence of the faster lateral decay in the plasma component than in thermal component. The plasma component manages to hold a positive phase under the pump beam but not away from it, which results in a negative signal wavelength.

Note that no outlier remains in the experimental data of ΔR_{ac}. In other words, whatever caused the unexpected behavior of R_{dc} for the sample at $P_{doping} = 4.62 \times 10^{18}\,\text{cm}^{-3}$ does not seem to impact ΔR_{ac}. This is confirmed by the quantitative study below.

[5] Attempts have been made to obtain better characterized fresh homogeneously doped substrates. However, NIST spreading resistance calibration standards [5] are no longer available and manufacturers seem to only sell batches of wafers and no single wafers. The samples of critical importance in this study (ultra-shallow junctions), however, have been fabricated during the study itself and properly monitored (surface oxide, substrate quality,...)

6.1.1.2 Quantitative Comparison

For a more convincing picture of the ability of the model to quantitatively predict the experimentally measured ΔR_{ac} on homogeneously doped silicon, we compare in Fig. 6.2 the experimental and theoretical variations with doping concentration in the four parameters of the damped wave of Eq. (6.3). ΔR_{ac} has been measured on 18 p-type doped samples and 2 n-type doped samples. For the sake of clarity, the theoretical curves are only shown for p-type doping since n-type doping would yield similar results (see Fig. 4.16). To facilitate the understanding, the theoretical curves for pure plasma and pure thermal signals are also shown.

The overall behavior of the signal on homogeneously doped substrates can be divided into three regimes.

First, at low doping ($<10^{19}\,\text{cm}^{-3}$), the pure plasma signal amplitude is approximately three times greater than the pure thermal signal amplitude. ΔR_{ac} is therefore plasma-dominated. The total signal amplitude shown in Fig. 6.2a is, however, always lower than the pure plasma signal amplitude due to the opposite sign of the non negligible thermal component. Interestingly, the signal amplitude shows a bump in the $10^{18}-10^{19}\,\text{cm}^{-3}$ doping range due to the non-monotonic behavior of the fundamental mode of the excess carrier distribution (Fig. 4.14c). The signal phase in Fig. 6.2c follows very closely the phase of the pure plasma signal. Finally, the behavior of the signal decay length and wavelength in Fig. 6.2b and d is more complex since they both depend not only on the ratios of the amplitudes of both signal components but also on the ratios of their decay lengths and wavelengths. As can be observed, the signal decay length follows the plasma decay length while always being longer. This is due to the slightly shorter decay length of the dominated thermal component. Due to the opposite signs of both signal components, the lateral decay in the dominated thermal component partially compensates the lateral decrease in the dominant plasma component, which results in a signal decay length longer than the plasma decay length. Similarly, the signal wavelength in Fig. 6.2d is always longer than the plasma wavelength (slower lateral change in phase) due to the shorter wavelength of the dominated thermal component.

Second, at high doping ($>10^{20}\,\text{cm}^{-3}$), the signal is thermally dominated. The signal phase follows the phase of the pure thermal signal while the amplitude is slightly lower than the pure thermal signal owing to the presence of the plasma component of opposite sign. The signal decay length is longer than the thermal decay length due to the short decay length of the plasma component. The signal wavelength is equal to the thermal wavelength since the plasma wavelength is infinite.

Finally, at intermediate doping concentrations, a transition region is observed where commensurate plasma and thermal components of opposite signs are in tight competition. This is by far the region where the signal behavior is most complex to understand and, therefore, to model. First, the signal amplitude decreases as long as it remains plasma-dominated (doping $<5\times10^{19}\,\text{cm}^{-3}$) and then rises once the pure thermal signal becomes dominant. The minimum value of the amplitude is not a zero since the two signal components are out of phase. Second, the signal phase

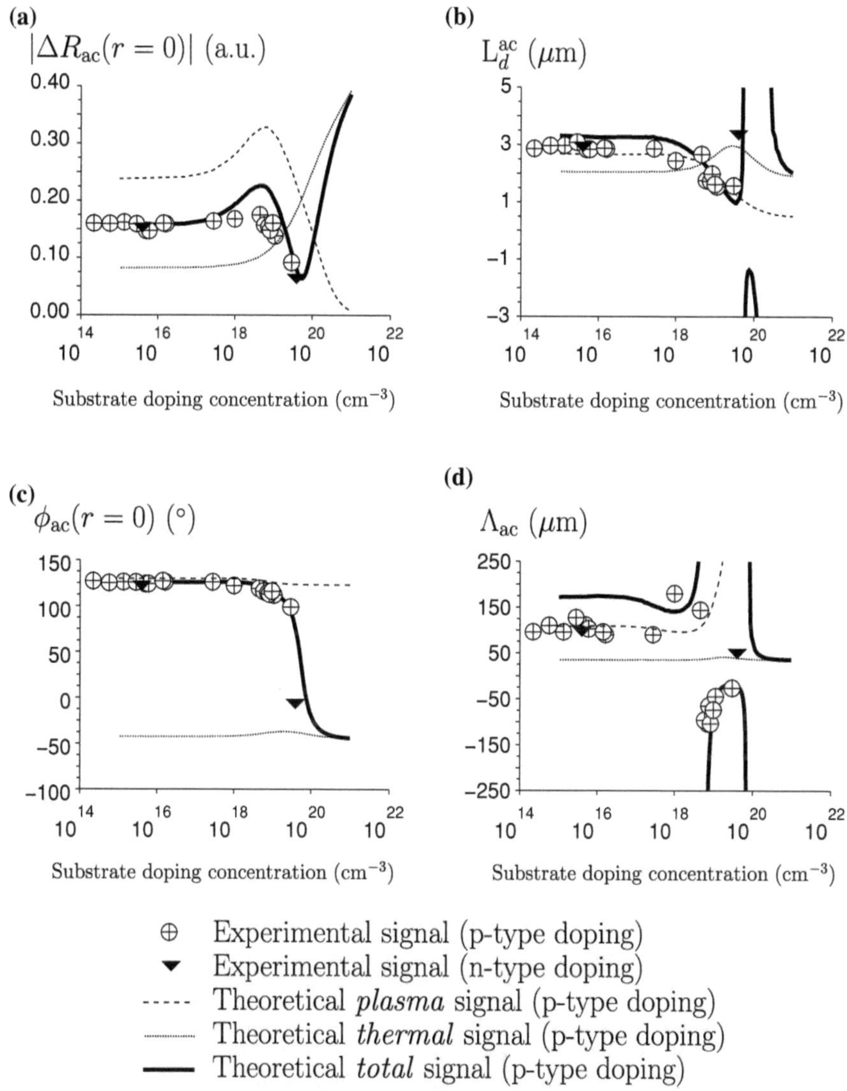

Fig. 6.2 Variation with doping concentration in **a** the amplitude at $r = 0$, **b** the signal decay length, **c** the phase at $r = 0$ and **d** the signal wavelength of the AC reflectance ΔR_{ac} on homogeneously doped substrates. The theoretically predicted signal values for the total signal on p-type doped substrates (*full lines*) are in good agreement with the experimental data on both n- and p-type substrates. The theoretical pure plasma signal (*dashed lines*) and pure thermal signal (*dotted lines*) are also indicated for completeness. The size of the experimental symbols is not representative of the error bar which is too small to be shown on the plotted scale (Appendix A.2)

undergoes a very abrupt transition approximately equal to $(180° + \phi_1^{pl} - \phi_1^{th})$. Finally, the signal decay length and wavelength show a very complicated behavior owing to their very subtle dependence on the ratios of the signal component amplitudes and decay lengths. In this transition region, both the signal decay length and wavelength can reach negative values and infinite values (asymptotes). In all generality, a negative signal decay length is obtained when *simultaneously* two signal components of opposite signs are commensurate *and* the lateral decay in the dominated component is much faster than the decay in the dominant component. More specifically, in the case of homogeneously doped substrates we are looking at, this situation only occurs for a narrow doping range around $10^{20}\,\mathrm{cm}^{-3}$ where the slightly dominant thermal component decays much more slowly than the dominated plasma component. Conversely, a negative signal wavelength is obtained when *simultaneously* two components of opposite signs are commensurate *and* the dominant signal component decays faster than the dominated component. This is observed for a relatively wide doping range around $10^{19}\,\mathrm{cm}^{-3}$ where the dominant plasma component decays faster than the dominated thermal component. An infinite signal decay length (resp. wavelength) occurs when the lateral decay (resp. lateral phase shift) in the dominant and dominated components perfectly compensate each other. The position of these asymptotes is even more complex since it further depends upon the phase difference and wavelength ratio of the two components.

Let us now look at the experimental data also presented in Fig. 6.2. Their agreement with the theoretical curves is excellent in all three regions, i.e. even in the very complex transition region. Notice, for example, the correctly predicted positions of the asymptotes of both characteristic lengths in Fig. 6.2b and d. Further, the bump on the signal amplitude predicted by our theory is observed experimentally, but it has a lower magnitude.

Though the agreement is good, it remains unsatisfactory in the low doping range, which is of the highest interest to this study. At low doping concentrations, both the experimental decay length and wavelength are closer to the pure plasma signal values than the total signal. This seems to indicate that our model slightly underestimates the ratio of the plasma component to the thermal component. We come back to this important observation in Sects. 6.2 and 6.3.

For the sake of completeness, note that an estimation of the normalization factor Γ_0 and phase θ_0 is needed to compare theory to experiments like in Fig. 6.2. In this figure, we take $\Gamma_0 = 4130$ and $\theta_0 = -57°$ so as to align the low doping values of the amplitude and phase. L_d^{ac} and Λ_{ac} are, however, independent from these factors.

6.2 Box-Like Doping Profile

Let us now summarize the various contributions to the measured signals in the case of a box-like doping profile. Just like for homogeneous doping, two effects are to be reported. First, the equilibrium carriers due to the box-like doping profile modify the steady refractive index profile (Chap. 3) and hence R_{dc} (Chap. 2). Second, as a consequence of the pump and probe lasers shining on the sample, an excess carrier distribution and an excess temperature distribution are generated. As shown in Chap. 5, these two distributions can easily be derived from the distributions generated in a homogeneously doped substrate, as found below the doped layer. In comparison with the homogeneous doping case of the previous section, the specificity of the present Section is that the *equilibrium and excess* carriers show an abrupt in-depth change in concentration at the junction depth, inducing a similar abrupt change in refractive index.

In other words, using Eqs. (2.17) and (3.16), R_{dc} can be written

$$R_{dc} = R_0 \left\{ 1 - \frac{4\beta}{n_0^2 - 1} \frac{1}{m_h} P_{doping} \left[1 - \cos(4\pi n_0 X_j / \lambda_{probe}) \right] \right\}$$

for p-type doped layers (6.9a)

$$R_{dc} = R_0 \left\{ 1 - \frac{4\beta}{n_0^2 - 1} \frac{1}{m_e} N_{doping} \left[1 - \cos(4\pi n_0 X_j / \lambda_{probe}) \right] \right\}$$

for n-type doped layers. (6.9b)

Equations (6.9a) and (6.9b) neglect the impact of the steady excess carrier and temperature distributions on R_{dc} (see e.g. Fig. 6.3a). They also assume low doping in the substrate. Finally, notice that, in agreement with formula (3.16) and the general conclusion of Chap. 3, we have neglected any possible variation in the optical constants due to BGN. As a consequence, the $\sin(4\pi n_0 X_j / \lambda_{probe})$ contribution to the signal vanishes. For further discussion of this approximation, we refer to Appendix A.6.

The AC reflectance, also assuming negligible impact of BGN on the optical constants, can be written

$$\begin{aligned}
\Delta R_{ac}(r) = {} & \frac{4R_0}{n_0^2 - 1} \Gamma_0 \exp(-i\theta_0) \\
& \times \Bigg\{ \underbrace{-\beta \left(\frac{1}{m_e} + \frac{1}{m_h} \right) \Delta N_{l1}(r) [1 - \cos(4\pi n_0 X_j / \lambda_{probe})]}_{\text{layer plasma component}} \\
& \underbrace{-\beta \left(\frac{1}{m_e} + \frac{1}{m_h} \right) \Delta N_{sub1}(r) \cos(4\pi n_0 X_j / \lambda_{probe})}_{\text{substrate plasma component}} \\
& + \underbrace{\delta \Delta T_1(r)}_{\text{thermal component}} \Bigg\} \\
= {} & \frac{4R_0}{n_0^2 - 1} \Gamma_0 \exp(-i\theta_0) \\
& \times \Bigg\{ \underbrace{-\beta \left(\frac{1}{m_e} + \frac{1}{m_h} \right) |\Delta N_{l1}(r=0)| \left[1 - \cos(4\pi n_0 X_j / \lambda_{probe}) \right] \exp[-i\phi_1^{pl}(r=0) - r/L_{l1}^{pl} - 2i\pi r/\Lambda_1^{pl}]}_{\text{layer plasma component}} \\
& \underbrace{-\beta \left(\frac{1}{m_e} + \frac{1}{m_h} \right) |\Delta N_{sub1}(r=0)| \cos(4\pi n_0 X_j / \lambda_{probe}) \exp[-i\phi_1^{pl}(r=0) - r/L_{d1}^{pl} - 2i\pi r/\Lambda_1^{pl}]}_{\text{substrate plasma component}}
\end{aligned}$$

$$+\underbrace{\delta|\Delta T_1(r=0)|\exp[-i\phi_1^{th}(r=0)-r/L_{d1}^{th}-2i\pi r/\Lambda_1^{th}]}_{\text{thermal component}}\Big\}. \tag{6.10}$$

Equations (6.9) and (6.10) are our model equations to explain the behavior of R_{dc} and ΔR_{ac} on box-like doping profiles. The addition of the excess carrier and temperature distributions obtained in Figs. 4.14 and 4.15 (only low doping region) as well as Fig. 5.4 complete our model. In summary, the DC reflectance R_{dc} shows sensitivity to both layer doping and junction depth. In particular, R_{dc} increases linearly with the doping concentration. As for the AC reflectance ΔR_{ac}, it is a three-component signal. The first component is the *layer plasma component*, related to the excess carrier concentration in the doped layer. This component always has the phase and wavelength of the plasma component on a homogeneous sample. The second component is the *substrate plasma component*, linked to the excess carrier concentration in the substrate. It has the wavelength of the plasma component on a homogeneous sample. However, due to the interference between the surface and interface modulated reflections, the sign of this component changes at a junction depth $X_j = \lambda_{probe}/(8n_0)$ ($\approx$22 nm). Its phase can therefore either be the same as on a homogeneous sample ($X_j < 22$ nm) or 180° out of phase ($X_j > 22$ nm). This remark is of the utmost importance for the general understanding of the signal behavior. The third and final component is the *thermal component* which is the same as the thermal component on a homogeneous silicon substrate. Note that, like R_{dc}, ΔR_{ac} is sensitive to layer doping and junction depth. However, only the layer plasma component of ΔR_{ac} is sensitive to doping. The sensitivity to doping consequently decreases with increasing doping (Fig. 5.4 or Eq. (5.10)). This explains why no correlation between ΔR_{ac} and dose or sheet resistance can be found. This is very different from the behavior of R_{dc}, which has shown very high correlation with sheet resistance [6]. This important difference between the behaviors of the two measured signals should be kept in mind for Chap. 7.

6.2.1 Comparison of the Model with Experimental Data

Similar to the case of homogeneously doped substrates, we separate the discussion into the qualitative behavior of experimental offset curves in Sect. 6.2.1.1 and the quantitative comparison with our model in Sect. 6.2.1.2.

6.2.1.1 Qualitative Comparison

Experimental offset curves measured on CVD box-like profiles are shown in Fig. 6.3 for two different p-type doping concentrations in the layer (CVD12 matrix, see Appendix A.3). First, looking at the DC reflectance (Fig. 6.3a), we clearly see that, at least qualitatively, it follows formulas (6.9) quite nicely. In particular, as assumed above, the curves are flat, which confirms that the time-independent excess carriers and temperature are not observed. It can also be seen that R_{dc} decreases when the depth of the box varies from 0 to approximately 44 nm ($= \lambda_{probe}/(4n_0)$). For deeper layers, R_{dc} increases with depth in agreement with the cosine behavior of formulas (6.9). As expected also, the amplitude of the variations in R_{dc} increases when the doping increases.

Figure 6.3b shows the offset curves of $|\Delta R_{ac}|$. When the beams are aligned, a non-monotonic behavior is observed when X_j varies, indicating the dominance of the plasma component over the thermal component. We will see in the next paragraphs that this behavior is a rectified cosine ($|\cos|$), as suggested by formula (6.10). Further, the amplitude of the cosine seems to increase when the layer doping increases. The behavior of the signal decay length, however, is difficult to determine based on this figure. Nonetheless, one clear observation is that the signal decay length is very close to that measured on a lowly doped substrate for very shallow X_j. The other, less clear, variations are discussed in the next section.

Figure 6.3c shows the offset curves of ϕ_{ac}. The phase for coincident beams also shows a non monotonic behavior. It is remarkable that, except for a couple of layers, the measured phase is either about 120° for depths below 30 nm or about −40° for deeper layers. This behavior is due to the sign change of the substrate plasma component at 22 nm. It can also be observed that, for low doping, the phase transition occurs for deeper layers. This transition can be explained from the competing behavior of the substrate and layer plasma components, the latter being larger at low doping (Fig. 5.4). Similarly, we see that the signal wavelength varies with depth but the behavior seems to be complex, just like that of the decay length. It can, however, be observed that the signal wavelength is approximately 120 μm when the phase is not in the transition region. This is about the wavelength of the plasma wave on a lowly doped substrate (see e.g. Fig. 4.14), which clearly indicates a plasma dominated behavior. In the transition region, negative wavelengths are observed. This is a consequence of the fast decay in layer plasma component which manages to hold the phase positive under the pump beam but not away from it.

6.2.1.2 Quantitative Comparison

Based on the previous observations, it seems that there is a good *qualitative* agreement between formulas (6.9) and (6.10) and Fig. 6.3. However, just like for homogeneously doped substrates, we would like to compare *quantitatively* the results of the developed model with the experimental data. For this purpose, we focus on the experimental data measured on the layers of the CVD12 matrix as above (layer doping $P_{doping} = 1.3 \times 10^{19}\,\text{cm}^{-3}$ and $6.4 \times 10^{19}\,\text{cm}^{-3}$), complemented with one set of the CVD13 matrix (layer doping $P_{doping} = 3.3 \times 10^{19}\,\text{cm}^{-3}$). The comparison between our theoretical predictions and the experimental data measured on these layers is shown in Fig. 6.4.

The quantitative discussion of the variations in R_{dc} is fairly simple. We show in Fig. 6.4a that this signal varies cosinusoidally with X_j. In particular, based on three different doping concentrations in the $1 - 6 \times 10^{19}\,\text{cm}^{-3}$ range, we show

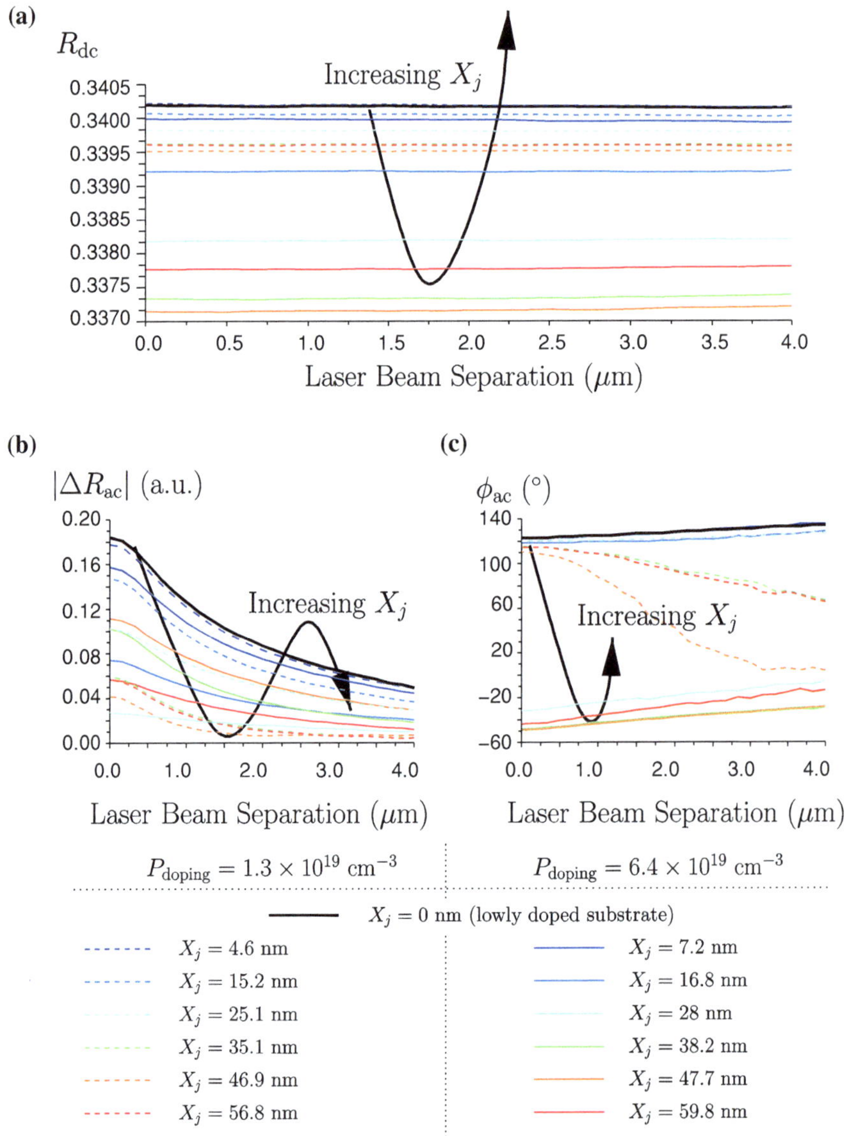

Fig. 6.3 Offset *curves*, i.e. variation with laser beam separation, of **a** the DC reflectance R_{dc}, **b** the amplitude and **c** phase of the AC reflectance ΔR_{ac} measured on p-type layers with various junction depths ranging from 0 to 60 nm and active doping concentrations of 1.3×10^{19} and 6.4×10^{19} cm^{-3} (values based on sheet resistance measurements with an assumed bulk mobility). All layers were grown on lowly doped device-quality substrates, the behavior of which is shown by the *thick black line*. The *arrows* show the trend observed when increasing junction depth

that the model gives very good agreement with experimental data. Notice, however, that the doping concentration used to fit the experimental R_{dc} is 75 % higher than the value based on sheet resistance measurements, independently from the doping concentration. This discrepancy is looked at in Sect. 6.3.1.

The quantitative discussion of ΔR_{ac} is more complex. The main complication lies in the very high sensitivity of ΔR_{ac} to the accuracy of the calculated substrate injection. On a box-like doping profile, ΔR_{ac} indeed depends on the excess carrier concentration in the substrate through two of its components, i.e. the substrate plasma component *and* layer plasma component. In other words, an error on the calculated excess carrier concentration in the substrate would propagate and lead to unacceptable results. It is therefore of critical importance that the excess carrier distribution in the substrate be correctly calculated. Yet, Sect. 6.1 shows that the agreement between the model and the measurements on homogeneously doped silicon, though good, is not 100 % quantitative. In particular, a possible small underestimation of the plasma component has been noted. For this reason, we propose a new calibration procedure providing an improved calculation of ΔN_{sub0}, ΔN_{sub1} and ΔT_1. Rather than using the distributions derived directly from our calculations, we propose to combine the insight offered by the developed theory and the experimental data measured on a lowly doped substrate. The proposed calibration procedure is as follows:

- we measure the experimental offset curve on a lowly doped substrate, such as the one the CVD layers have been grown on (e.g. before growth of the layer)
- we fit Eq. (6.8) and the three-dimensional linear ambipolar and heat diffusion equations (Eqs. (4.60b), (4.60d)) to the measured curves with the uniform ambipolar diffusivity and the uniform carrier lifetime as fitting parameters. The fitting algorithm actually also involves the normalization factor Γ_0 and phase θ_0. Importantly, this fitting procedure gives a unique solution. The slopes of the measured curves indeed give access to the ambipolar diffusivity and the carrier lifetime (see e.g. the asymptotic behavior of L_{d1}^{pl} and Λ_1^{pl} in Fig. 4.10). Further, Γ_0 and θ_0 are fixed by the values of the signals themselves.
- ΔN_{sub1} and ΔT_1 are direct outputs of the fitting procedure. Note that the obtained distributions are very close to the distributions calculated in Chap. 4. Typically, on a lowly doped substrate, a 2×10^{-7} s carrier lifetime is obtained, in good agreement with Auger recombinations at the expected injection. The obtained ambipolar diffusivity, however, is about $4\,\text{cm}^2\,\text{s}^{-1}$, i.e. about two times smaller than the theoretical values (see e.g. Fig. 4.5). As a consequence, the obtained $|\Delta N_{sub1}|$ is about 30% higher than calculated, while the temperature is roughly unchanged. The decay lengths, wavelengths and phases of both components remain approximately unchanged ($\leq$5% difference). This discrepancy is looked at in Sect. 6.3.2.1.
- ΔN_{sub0} is calculated based on the steady linear ambipolar diffusion equation (similar to Eq. (4.60a) with the consideration of the probe laser), assuming the uniform ambipolar diffusivity and carrier lifetime derived in this procedure. The obtained ΔN_{sub0} on a lowly doped substrate is roughly 30 % higher than the calculated one due to the lower ambipolar diffusivity. The decay length, wavelength

and phase are approximately the same as their calculated counterparts ($\leq$5 % difference).

Notice that, besides providing a simple and direct access to ΔN_{sub0}, ΔN_{sub1} and ΔT_1, this calibration procedure shows some other great advantages. First, values for the normalization factor Γ_0 and phase θ_0 are derived consistently. These values would have had to be assumed otherwise (see e.g. Sect. 6.1 for homogeneous substrates). Second, the procedure can be generalized to other substrates, e.g. with preamorphization. This proves very useful in Sect. 7.2. Finally, given that D_a is measured rather than calculated, the modeling error is also reduced (Sect. 6.3.2.1). The calibration procedure is, however, still perfectible. First, it assumes a fixed value for the absorption coefficients and irradiances of both lasers. Any actual variation in any of these coefficients during the measurements (e.g. plasma-induced BGN, long-term instability of the laser irradiances,...) would therefore mistakenly be attributed to variations in Γ_0. Second, it assumes uniform values of D_a and τ. We will see in Sect. 7.2 that this assumption is quite detrimental in the case of preamorphized substrates. We, however, still believe this new calibration procedure is a great improvement over the previously used procedure, which would attribute all the possible variations to changes in the measured profiles.

The results based on our model Eq. (6.10) and on the proposed calibration procedure are shown in Fig. 6.4b–e for three doping concentrations in the $1-6\times10^{19}\,\text{cm}^{-3}$ range. The values of the normalization factor and phase obtained by fitting of the substrate offset curves are $\Gamma_0 = 2857$ (a.u.) and $\theta_0 = -61°$. We separate the discussion of the signal behavior into three. First, we focus on highly doped layers (red curves). We then proceed to the case of moderately doped layers (black curves) and finish with the lowly doped layers (blue curves).

Let us first discuss the case of a high doping concentration in the layer (red curves). The signal amplitude shown in Fig. 6.4b follows a rectified cosine ($|\cos|$) behavior, which indicates a dominant substrate plasma component. The amplitude reaches a minimum at $X_j^{\text{transition}} \approx 22\,\text{nm}$ as a result of the sign change of the substrate plasma component. More precisely, $X_j^{\text{transition}}$ corresponds to the depth where the substrate plasma component and layer plasma component are equal but of opposite signs (if the thermal component is neglected) and is therefore deeper than 22 nm (25 nm in this highly doped case). An approximate expression of $X_j^{\text{transition}}$ can be found by solving

$$\cos(4\pi n_0 X_j^{\text{transition}}/\lambda_{\text{probe}}) = \frac{-|\Delta N_{l1}/\Delta N_{sub1}|}{1-|\Delta N_{l1}/\Delta N_{sub1}|}. \tag{6.11}$$

Note already that the signal behavior around $X_j^{\text{transition}}$ is very complex since all three signal components are roughly equal. Moving on to the signal phase, shown in Fig. 6.4d, it is first very close to ϕ_1^{pl} for $X_j < X_j^{\text{transition}}$ and then undergoes an abrupt transition at $X_j^{\text{transition}}$ due to the sign change of the substrate plasma component. The total phase variation is almost equal to 180°, but is slightly smaller due

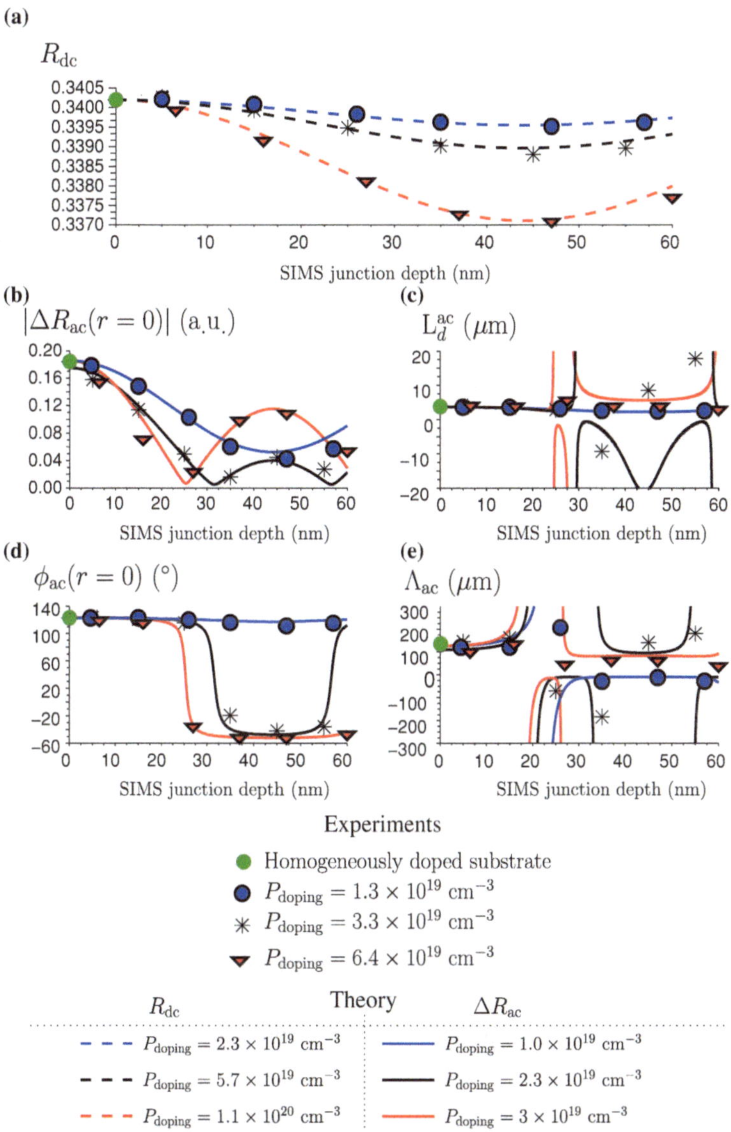

Fig. 6.4 Variation with junction depth in **a** the DC reflectance R_{dc}, **b** the amplitude at $r = 0$, **c** the signal decay length, **d** the phase at $r = 0$ and **e** the signal wavelength of the AC reflectance ΔR_{ac} measured on p-type doped layers with doping concentrations 1.3×10^{19} cm^{-3} (*blue circles*), 3.3×10^{19} cm^{-3} (*black asterisks*) and 6.4×10^{19} cm^{-3} (*red triangles*). These active doping concentration values are based on sheet resistance measurements and an assumed bulk mobility. The theoretically predicted signal values for the total signal on p-type doped substrates (*colored lines*) are in good agreement with the experimental data though the active doping concentrations shown do not match the experimental values. The active doping concentrations for the theoretical values of R_{dc} and ΔR_{ac} also are different. The size of the experimental symbols is not representative of the error bar which is too small to be shown on the plotted scale (Appendix A.2)

to the different phase of the thermal component. As can be expected, the discussion of the signal decay length and wavelength shown respectively in Fig. 6.4c and d is very complex owing to the presence of three signal components. We therefore keep the discussion simple and focus on the possibility for negative values. Starting with the signal decay length, the conditions for a negative signal decay length are found in a narrow X_j range deeper than $X_j^{\text{transition}}$. In this region, indeed, the substrate and layer plasma components are commensurate and of opposite signs and, *simultaneously*, the dominated layer plasma component decays faster than the dominating substrate plasma component. This explains the negative signal decay length when $X_j \geq X_j^{\text{transition}}$. A similar argument holds for the negative signal wavelength. For junction depths included approximately between 22 nm and $X_j^{\text{transition}}$, the substrate and layer plasma components are commensurate and, *simultaneously*, the dominant layer plasma component decays faster than the dominated substrate plasma component. As a result, the signal wavelength is negative in this narrow X_j range.

Second, for layers with an intermediate active doping concentration (black curves), the behavior is similar. The observed differences with respect to the red curves (higher doping) are due to the larger ΔN_l and, hence, the higher amplitude of the layer plasma component. First, the transition region moves towards deeper X_j, i.e. $X_j^{\text{transition}} \approx 32$ nm. Second, the X_j ranges where a negative signal decay length or wavelength can be found widen dramatically.

Finally, the lowly doped layers (blue curves) show quite a different behavior. The reason for this difference is that $|\Delta N_{l1}/\Delta N_{sub1}| > 0.5$. In this case, Eq. (6.11) predicts that no transition should occur since the layer plasma component is larger than the substrate plasma component independently from X_j. This is indeed what is observed. As a consequence, the phase and decay length remain approximately equal to respectively ϕ_1^{pl} and L_{d1}^{pl} independently from X_j. Note, however, that the dominance of the layer plasma component independently from X_j allows a negative signal wavelength on a very broad X_j range.

This overall theoretical behavior is in very good agreement with the experimental data. In particular, Fig. 6.4b and d show the excellent agreement between the measured and theoretical values of the amplitude $|\Delta R_{\text{ac}}(r = 0)|$ and phase $|\phi_{\text{ac}}(r = 0)|$ for coincident beams. The agreement between calculated and experimental signal decay lengths L_d^{ac} and wavelengths Λ_{ac} respectively in Fig. 6.4c and e is also satisfactory.

Note that most of the observed discrepancies between experimental and theoretical L_d^{ac} and Λ_{ac} are just a consequence of the definition of these characteristic lengths. It has to be reminded that, though ΔR_{ac} has been defined as a damped wave in Eq. (6.3), it is actually the superposition of three of them (Eq. (6.10)). It can therefore be expected that, in order to have well-defined signal decay lengths and wavelengths, one or, at most, two signal components must be dominant. This is the case on homogeneously doped substrates, where a very good agreement has indeed been found between the theoretical and experimental values of these characteristic lengths. Very importantly, Fig. 7.5c and the related discussion in Sect. 7.1.2 of the next chapter confirm that our model accounts for the lateral behavior of ΔR_{ac} with great accuracy.

As a consequence of the observed good agreement, the model Eq. (6.10) can therefore be believed to quantitatively model the variations in ΔR_{ac}. However, it can be noticed that the active doping concentrations needed to obtain the theoretical curves of Fig. 6.4 are different from the ones derived by sheet resistance measurements. They are also different from those needed for R_{dc}. More particularly, the sensitivity of ΔR_{ac} to doping seems to be overestimated by the model. In Sect. 6.3.2.2 below, we determine the possible explanations for the disagreement.

As a final remark, note that Appendix A.5 studies qualitatively the behavior of ΔR_{ac} on n-type layers and shows that they behave similarly to p-type layers. They do not directly appear in this chapter because, unfortunately, these layers were grown in the early phase of this project. As a consequence, it is impossible to study them quantitatively. We indeed have no measurement on the homogeneous substrate before growth of the CVD layer and can therefore not apply our calibration procedure. Further, these n-layers have been grown on substrates of low quality (monitor wafers). This proves to strongly reduce the repeatability of the technique as already mentioned and as shown in Appendix A.2.

6.3 Discussion of the Modeling Error

As confirmed in the previous sections of the present chapter, our model equations give very satisfactory agreement with experimental data. It cannot be claimed, however, that the model is perfectly quantitative. Some discrepancies between experiments and theory have indeed been observed both on the DC and AC reflectances. We would like to investigate the most likely causes for the disagreements. Note that, in spite of the important conclusions drawn in this section, the model will remain as discussed thus far. This section therefore must be considered as an open discussion. It also contains some of our recommendations for future work and for possible improvements of the model.

We separate this study into two items. First, we look at the discrepancies observed on R_{dc} in Sect. 6.3.1. Section 6.3.2 then focuses on the error on ΔR_{ac}.

6.3.1 Modeling Error on R_{dc}

Figure 6.4 shows that our model Eqs. (6.9a) and (6.9b) correctly predict the behavior of R_{dc} on box-like profiles. Nonetheless, we have noticed that our model only fits the experimental behavior of R_{dc} if we use doping concentrations 75 % higher than expected from accurate (zero-penetration) sheet resistance measurements. Two causes are likely to explain this discrepancy.

First, the difference could be attributed to a possibly underestimated electrorefractive effect in our model. As already mentioned in Chap. 3, this underestimation is likely since large variations in the quantification of the electrorefractive effect have been published [7] (up to a factor 10 variation).

Second, the observed difference can equally come from an error on the active doping concentration derived from the sheet resistance measurements. As already mentioned in Sect. 1.2.1.3, sheet resistance measurements give access to the conductivity in the layer, not its active doping concentration. In other words, a value of the majority carrier mobility is needed to access the active doping concentration. We have made use of the bulk mobility values given by Ref. [8]. It is, however, well-known that mobility in a doped layer is degraded (up to 30 %) by the presence of inactive dopants [9]. The derived active doping concentration, which we have here called "experimental", should therefore only be taken as a lower limit of the active doping concentration.

The discrepancy observed on R_{dc} is likely to be due to both error sources. It is unfortunately not possible to determine which, if any, is dominant. To solve this problem, we recommend to run mobility measurements on the full matrix of p-type CVD layers, e.g. with a Hall-based technique [10, 11]. Knowing the actual values of the majority carrier mobility in the layer, the correct active doping concentrations will be found. As a consequence, an experimental value for $\partial n/\partial P$ will be derived. We also recommend to run the same experiment on n-type layers so as to determine an experimental value for $\partial n/\partial N$. This is obviously needed for the quantitative understanding of R_{dc} on n-type layers. As we show below, it is also needed so as to improve the modeling of ΔR_{ac} on both p- and n-type layers.

6.3.2 Modeling Error on ΔR_{ac}

Figures 6.2 and 6.4 show that our model Eqs. (6.8) and (6.10) correctly predict the behavior of ΔR_{ac} respectively on homogeneously doped substrates and on box-like profiles. However, the quantitative comparison between experiments and the results of our model highlights two main issues. First, measurements on lowly doped substrates underline the underestimation of the ratio of the plasma component to the thermal component. We study the possible sources for such an error in Sect. 6.3.2.1. Second, the modeling of the dependence of ΔR_{ac} upon doping needs to be improved. In particular, we have noticed an overestimated sensitivity to doping. The origin of this effect is discussed in Sect. 6.3.2.2.

6.3.2.1 Underestimated Plasma-to-Thermal Component Ratio

The experimental data on lowly doped substrates show that the ratio of the plasma component to the thermal component is slightly underestimated by the model (see signal decay length and wavelength in Fig. 6.2). Three possible error sources can be considered.

The first error source is the possible underestimation of the electrorefractive effect. This effect has also been mentioned as a likely source of error in the modeling of R_{dc}.

The fact that this effect is observable on both R_{dc} and ΔR_{ac} cannot be a coincidence and is considered as a confirmation of the underestimated electrorefractive effect. As a consequence, the problem should, at least partially, be solved when experimental values for $\partial n/\partial P$ and $\partial n/\partial N$ are measured, as we suggest in Sect. 6.3.1 above.

The second possible source of error lies in the underestimated excess carrier concentration. Two arguments are in favor of the underestimation. First, we have mentioned in Sect. 4.1.3.1 that our model underestimates the band-to-band absorption coefficient α_{BTB}. We have also shown how much this parameter can affect the excess carrier concentration (Fig. 4.14). Second, we have seen that, in order to fit the slopes of the offset curves measured on homogeneously doped substrates, a reduced ambipolar diffusivity is needed (Sect. 6.2.1), leading to a 30 % increase in excess carrier concentration. Physically speaking, this reduction in ambipolar diffusivity is in contradiction with the excellent agreement between the theoretical and experimental D_a values shown in Fig. 4.5. One way of having theory and experiments to agree is to *speculate* that the BGN counter-diffusive terms D_n^{BGN} and D_p^{BGN} start to impact the ambipolar diffusivity significantly at excess carrier concentrations of the order of $10^{18}\,\mathrm{cm}^{-3}$, i.e. for larger injections than the experimental data of Fig. 4.5. In that case, at high injection, their negative contribution D_a^{BGN} to the ambipolar diffusivity

$$\begin{aligned} D_a^{\mathrm{BGN}} &= \frac{\mu_p D_n^{BGN} + \mu_n D_p^{BGN}}{\mu_n + \mu_p} \\ &= -\frac{\mu_n \mu_p}{\mu_n + \mu_p} \Delta N \frac{1}{q} \left| \frac{\partial E_g}{\partial \Delta N} \right| . \end{aligned} \tag{6.12}$$

would lead to a reduced ambipolar diffusivity. This, in turn, would increase the excess carrier concentration and, therefore, generate a positive feedback loop. In any case, both the underestimated α_{BTB} and overestimated D_a point towards the importance of an accurate quantification of the plasma-induced BGN in our study. Note that plasma-induced BGN can be accounted for by the model we use [12] but it has not been checked experimentally (only doping-induced BGN). We therefore believe this problem would be partially solved with a better quantified plasma-induced BGN model.

Interestingly, the underestimated carrier injection can also serve as a justification for an assumption which we have made in this thesis. We have assumed a zero surface recombination velocity (SRV) at all boundaries. The initial reason is that this effect cannot be modeled quantitatively [13]. We now can see that this assumption is acceptable since adding a nonzero SRV at any boundary would have led to a further reduction in carrier injection. We believe the negligible values of the SRV in this work can be partially explained by the high modulation frequency of TP. A diffusion-limited behavior indeed shows little sensitivity to recombinations (see e.g. Sect. 4.2.1). Another explanation lies in our measurement procedure (Appendix A.1.1) where the signals are recorded only about 25 s after starting shining the pump and probe lasers on the sample surface. This indeed tends to charge the oxide and therefore to reduce the SRV [2].

As for the third source of error, i.e. the possible underestimation of the excess temperature of thermorefractive effect, we do not have any evidence for it. We therefore suggest this effect should be looked at when the above-mentioned two problems are solved.

6.3.2.2 Overestimation of the Sensitivity of ΔR_{ac} to Layer Doping

The active doping concentrations needed to fit the experimental behavior of ΔR_{ac} on box-like profiles in Fig. 6.4 do not match the active doping concentrations derived from sheet resistance measurements. In particular, it has been observed that the sensitivity of ΔR_{ac} to doping seems to be overestimated. This sensitivity can be calculated from the following derivative

$$\begin{aligned} \left|\frac{\partial \Delta R_{ac}}{\partial P_{doping}}\right| &= \left|\frac{\partial \Delta R_{ac}}{\partial \Delta N_{l1}}\right| \left|\frac{\partial \Delta N_{l1}}{\partial P_{doping}}\right| \\ &= \frac{4R_0}{n_0^2 - 1}\Gamma_0 \beta \left(\frac{1}{m_e} + \frac{1}{m_h}\right) \left|1 - \cos(4\pi n_0 X_j/\lambda_{probe})\right| \left|\frac{\partial \Delta N_{l1}}{\partial P_{doping}}\right|. \end{aligned} \tag{6.13}$$

The previous considerations of the modeling error in this Section have implied that the Drude β factor and the electrorefractive effect it embodies are actually underestimated. The error would therefore originate from an overestimated $|\partial \Delta N_{l1}/\partial P_{doping}|$. In cases of low injection (high doping) in the doped layers, we have, using Eq. (5.10)

$$\begin{aligned} \left|\frac{\partial \Delta N_{l1}}{\partial P_{doping}}\right| &= \left|\frac{\partial}{\partial P_{doping}}\left[\frac{2\Delta N_{sub0}\Delta N_{sub1}}{P_{doping}} \exp\left(-\frac{E_g^l - E_g^{sub}}{k_b T}\right)\right]\right| \\ &= \frac{2\Delta N_{sub0}|\Delta N_{sub1}|}{P_{doping}} \exp\left(-\frac{E_g^l - E_g^{sub}}{k_b T}\right)\left|\frac{1}{k_b T}\frac{\partial E_g^l}{\partial P_{doping}} + \frac{1}{P_{doping}}\right|. \end{aligned} \tag{6.14}$$

Apparent from Eq. (6.14) is that an overestimated sensitivity to doping is likely to come from overestimated values of ΔN_{sub0}, i.e. of the steady carrier injection level below the doped layer. We believe this is a direct consequence of our supposed independence of the substrate injection from the doped layer. As can be seen in Fig. 5.3, though the fundamental mode of the excess carrier distribution in the substrate is expected to vary little with the doped layer, its constant mode decays quite strongly for highly doped layers ($\approx$20 % for a layer doping concentration of $2 \times 10^{20}\,\text{cm}^{-3}$). To enhance the accuracy of the model, it is therefore critical that this effect be taken into account. Two solutions can be considered. First, on the theoretical side, one should go back to the full time-dependent (or steady periodic) drift-diffusion simulations. As already mentioned, however, both the implementation and resolution of this problem are complex and time-consuming tasks. An alternative, experimental, solution is to increase the pump irradiance and, thus, reduce the impact of the depletion effect observed in Fig. 5.3. We obviously recommend the latter experimental solution. It is very interesting to note that, as can be seen in this section, ΔN_{sub0} is overestimated while, as shown in the previous section, ΔN_{sub1} is underestimated.

Further, plasma- and doping-induced BGN effects also appear in Eq. (6.14) via respectively E_g^{sub} and E_g^l. They can therefore also be identified as a source of error. Their presence as argument of an exponential shows once again that very high accuracy is needed for the BGN model.

Note, finally, that the presence of inactive dopants in the doped layers could also explain partly the observed overestimated substrate injection [14]. However, the high activation of our CVD layers (≥50 %) lets us think that this effect is marginal.

6.4 Summary

In this chapter, we have derived model equations to explain the experimental behavior of the DC reflectance R_{dc} and AC reflectance ΔR_{ac}. These equations read respectively for a homogeneously doped substrate

$$R_{\mathrm{dc}}^{\mathrm{Substrate}} = R_0 \left[1 - \frac{4\beta}{m_h(n_0^2 - 1)} P_{\mathrm{doping}} \right] \text{ for p-type doping} \tag{6.15a}$$

$$R_{\mathrm{dc}}^{\mathrm{Substrate}} = R_0 \left[1 - \frac{4\beta}{m_e(n_0^2 - 1)} N_{\mathrm{doping}} \right] \text{ for n-type doping} \tag{6.15b}$$

$$\Delta R_{\mathrm{ac}}^{\mathrm{Substrate}}(r) = \frac{4R_0}{n_0^2 - 1} \Gamma_0 \exp(-i\theta_0) \left[-\beta \left(\frac{1}{m_e} + \frac{1}{m_h} \right) \Delta N_1(r) + \delta \Delta T_1(r) \right], \tag{6.15c}$$

and for a box-like profile

$$R_{\mathrm{dc}}^{\mathrm{Box}} = R_0 \left\{ 1 - \frac{4\beta}{n_0^2 - 1} \frac{1}{m_h} P_{\mathrm{doping}} \left[1 - \cos(4\pi n_0 X_j / \lambda_{\mathrm{probe}}) \right] \right\} \quad \text{for p-type doped layers} \tag{6.16a}$$

$$R_{\mathrm{dc}}^{\mathrm{Box}} = R_0 \left\{ 1 - \frac{4\beta}{n_0^2 - 1} \frac{1}{m_e} N_{\mathrm{doping}} \left[1 - \cos(4\pi n_0 X_j / \lambda_{\mathrm{probe}}) \right] \right\} \quad \text{for n-type doped layers} \tag{6.16b}$$

$$\Delta R_{\mathrm{ac}}^{\mathrm{Box}}(r) = \frac{4R_0}{n_0^2 - 1} \Gamma_0 \exp(-i\theta_0) \times \left\{ -\beta \left(\frac{1}{m_e} + \frac{1}{m_h} \right) \Delta N_{l1}(r) [1 - \cos(4\pi n_0 X_j / \lambda_{\mathrm{probe}})] \right.$$

$$- \beta \left(\frac{1}{m_e} + \frac{1}{m_h} \right) \Delta N_{sub1}(r) \cos(4\pi n_0 X_j / \lambda_{\text{probe}}) + \delta \Delta T_1(r) \Big\}. \tag{6.16c}$$

These equations, together with Figs. 4.14 (red curves), 4.15 (red curves) and 5.4 constitute our model. The model has been tested on homogeneously doped substrates and p-type CVD layers. Though the model is in good agreement with experimental data, the doping dependence of both R_{dc} and ΔR_{ac} still requires some improvement. Some suggestions and directions have been given to refine the model.

To conclude, let us write our model equations for an active doping profile with an arbitrary shape

$$R_{\text{dc}}^{\text{Profile}} = R_0 \left\{ 1 - \frac{4\beta}{n_0^2 - 1} \frac{1}{m_h} \left[P_{\text{doping}}(z=0) + \int_{0+}^{+\infty} \frac{\partial P_{\text{doping}}(z)}{\partial z} \cos(4\pi n_0 z / \lambda_{\text{probe}}) dz \right] \right\}$$

for p-type active doping profiles (6.17a)

$$R_{\text{dc}}^{\text{Profile}} = R_0 \left\{ 1 - \frac{4\beta}{n_0^2 - 1} \frac{1}{m_h} \left[N_{\text{doping}}(z=0) + \int_{0+}^{+\infty} \frac{\partial N_{\text{doping}}(z)}{\partial z} \cos(4\pi n_0 z / \lambda_{\text{probe}}) dz \right] \right\}$$

for n-type active doping profiles (6.17b)

$$\begin{aligned} \Delta R_{\text{ac}}^{\text{Profile}}(r) &= \frac{4R_0}{n_0^2 - 1} \Gamma_0 \exp(-i\theta_0) \\ &\times \Big\{ - \beta \left(\frac{1}{m_e} + \frac{1}{m_h} \right) \left[\Delta N_{l1}(r, z=0) + \int_{0+}^{+\infty} \frac{\partial \Delta N_{l1}(r, z)}{\partial z} \cos(4\pi n_0 z / \lambda_{\text{probe}}) dz \right] \\ &+ \delta \Delta T_1(r) \Big\}. \end{aligned} \tag{6.17c}$$

which, together with Eq. (5.8), Figs. 4.14 and 4.15, constitute our model to explain the behavior of the DC and AC reflectances on active doping profiles of arbitrary shape.

References

1. J. Opsal, M.W. Taylor, W.L. Smith, A. Rosencwaig, Temporal behavior of modulated optical reflectance in silicon. J. Appl. Phys. **61**(1), 240–248 (1987)
2. F. Dortu, Low-Frequency Modulated Optical Reflectance for the One-Dimensional Characterization of Ultra-Shallow Junctions. Ph.D. thesis, Katholieke Universiteit Leuven, 2009
3. A. Mandelis, L. Nicolaides, Y. Chen, Structure and the reflectionless/refractionless nature of parabolic diffusion-wave fields. Phys. Rev. Lett. **87**(2), 020801/1-4 (2001)
4. J. Bogdanowicz, F. Dortu, T. Clarysse, W. Vandervorst, E. Rosseel, N.D. Nguyen, D. Shaughnessy, A. Salnik, L. Nicolaides, Nondestructive extraction of junction depths of active doping profiles from photomodulated optical reflectance offset curves. J. Vac. Sci. Technol. B **28**(1), C1C1–C1C7 (2010)
5. http://www.nist.gov/srm/

6. E. Rosseel, J. Bogdanowicz, T. Clarysse, W. Vandervorst, C. Ortolland, T. Hoffmann, A. Salnik, L. Nicolaides, S.H. Han, D.H. Petersen, R. Lin, O. Hansen, Study of submelt laser induced junction nonuniformities using therma-probe. J. Vac. Sci. Technol. B **28**(1), C1C21–C1C26 (2010)
7. R. Thalhammer, Internal Laser Probing Techniques for Power Devices: Analysis, Modeling and Simulation. Ph.D. thesis, Universitat Muchen, 2000, http://tumb1.biblio.tu-muenchen.de/publ/diss/ei/2000/thalhammer.pdf
8. D.B.M. Klaassen, A unified mobility model for device simulation. 1. model-equations and concentration-dependence. Solid-State Electron. **35**(7), 953–959 (July 1992)
9. T. Clarysse, J. Bogdanowicz, J. Goossens, A. Moussa, E. Rosseel, W. Vandervorst, D.H. Petersen, R. Lin, P.F. Nielsen, O. Hansen, G. Merklin, N.S. Nennett, N.E.B. Cowern, On the analysis of the activation mechanisms of sub-melt laser anneals. Mater. Sci. Eng. B **154–155**, 24–30 (2008)
10. D.K. Schroder, *Semiconductor Material and Device Characterization, Chap. 9* (Wiley-IEEE Press, New York, 2006)
11. D.H. Petersen, O. Hansen, R. Lin, P.F. Nielsen, Micro-four-point probe hall effect measurement method. J. Appl. Phys. **104**(1), 013710-1-10 (2008)
12. A. Schenk, Finite-temperature full random-phase approximation model of band gap narrowing for silicon device simulation. J. Appl. Phys. **84**(7), 3684–3695 (1998)
13. F. Dortu, J. Bogdanowicz, T. Clarysse, W. Vandervorst, Impact of band gap narrowing and surface recombination on photoelectrothermal modulated optical reflectance power curves. J. Vac. Sci. **26**, 322–332 (2008)
14. J. Bogdanowicz, F. Dortu, T. Clarysse, W. Vandervorst, D. Shaughnessy, A. Salnik, L. Nicolaides, Impact of inactive dopants in chemical vapor deposition layers on photomodulated optical reflectance. Mater. Sci. Eng. B **154–155**, 234–239 (2008)

Chapter 7
Application of the Model to Carrier Profiling

In the previous chapters, we have developed, tested and commented a physical model to explain the general behavior of the DC reflectance R_{dc} and AC reflectance ΔR_{ac}, as measured with TP on homogeneously and non-homogeneously doped silicon. In this chapter, the insights of the previous chapters are put together in order to solve the inverse problem. In other words, in this chapter, we show how to deduce information about unknown active doping profiles based on TP measurements. We start, in Sect. 7.1, by showing that it is possible to determine the junction depth of a box-like doping profile based on ΔR_{ac}. This signal has indeed shown a strong sensitivity to junction depth (Eq. (6.10)). Using ΔR_{ac} is a very promising approach for the non-destructive determination of the junction depth of an unknown doping profile. It is of particular interest to mention that, though modeling is involved in the development of the X_j determination technique, the derivation of the X_j value itself is based on processing of experimental data only. This is of course a great advantage since we have shown that the layer-doping dependence of ΔR_{ac} is yet to be better quantified. Further, since the ultimate objective of this work lies in the reconstruction of complete active doping profiles, we develop, in Sect. 7.2, a model-based method for full profile characterization by combining the use of ΔR_{ac} offset curves and R_{dc}. We will see that this technique shows a promising agreement with other carrier and dopant profiling techniques, with the great benefit of being non-destructive, local and fast. We will also underline the current limitations of the technique.

Interestingly, in the course of this chapter, we will find some additional confirmation of the ability of our model equations to explain the qualitative and quantitative behavior of R_{dc} and ΔR_{ac}.

7.1 Model-Free Determination of Junction Depths

As clearly highlighted in formula (6.10), ΔR_{ac} is very sensitive to X_j. However, though this signal varies also with the doping concentration in the layer, the sensitivity to the latter drops at the high doping concentrations needed to make efficient

J. Bogdanowicz, *Photomodulated Optical Reflectance*, Springer Theses,
DOI: 10.1007/978-3-642-30108-7_7,

USJs ($>10^{20}$ cm^{-3}). In other words, ΔR_{ac} is certainly a very promising signal to help determine X_j but, on its own, is quite unlikely to be useful for full profile reconstruction. It also has to be recalled that the variations in ΔR_{ac} with layer doping are not accounted for with sufficient accuracy by our model (Chap. 6). In this section, we therefore focus on the use of ΔR_{ac} for the extraction of X_j.

In order to isolate X_j in the expression of ΔR_{ac}, we somehow have to cancel the other dependencies of ΔR_{ac} and thus the layer plasma and thermal components. In order to reach that goal, we first propose to use offset curves in Sect. 7.1.1. We show that it is, in theory, possible to derive an *absolute* value of the junction depth of a box-like doping profile. Second, in Sect. 7.1.2, we develop an alternative technique for the *relative* determination of X_j. Advantageously, these two techniques do not involve any modeling in the X_j determination, at the expense of one or two calibration measurements.

7.1.1 Absolute Determination of Junction Depths

The main requirement for the determination of X_j from ΔR_{ac} measurement is to eliminate the layer plasma component as well as the thermal component. The solution investigated in this section is the separation of the laser beams, i.e. offset curves. Based on the decay lengths observed in Figs. 4.14d, 4.15d (red curves) and 5.4, it is obvious that the three signal components of Eq. (6.10) do not decay laterally at the same pace. In particular, our calculations show that the substrate plasma component has the largest decay length (≈2.7 μm). In other words, for sufficiently separated beams, the substrate plasma component is the only remaining signal component, i.e.

$$\lim_{r\to\infty} \Delta R_{ac}^{Box}(r) = -\frac{4R_0}{n_0^2-1}\Gamma_0 \exp(-i\theta_0)\beta\left(\frac{1}{m_e}+\frac{1}{m_h}\right)\Delta N_{sub1}(r) \times \cos(4\pi n_0 X_j/\lambda_{probe}), \quad (7.1)$$

where X_j is now the only parameter linked to the doping profile. Similarly on a homogeneously doped substrate, we also have

$$\lim_{r\to\infty} \Delta R_{ac}^{Substrate}(r) = -\frac{4R_0}{n_0^2-1}\Gamma_0 \exp(-i\theta_0)\beta\left(\frac{1}{m_e}+\frac{1}{m_h}\right)\Delta N_{sub1}(r). \quad (7.2)$$

Since the fundamental mode of the excess carrier distribution in the substrate is independent from the doped layer,[1] we have, when taking the ratio of formulas (7.1) and (7.2),

[1] We believe it is important to remind that ΔN_{sub0} is expected to vary with the layer doping, ΔN_{sub1} is not (Fig. 5.3).

$$\lim_{r\to\infty} \frac{\Delta R_{ac}^{Box}(r)}{\Delta R_{ac}^{Substrate}(r)} = \cos(4\pi n_0 X_j/\lambda_{probe}). \tag{7.3}$$

The meaning of Eq. (7.3) is that the ratio of the signals measured respectively on the studied sample and on a lowly doped substrate such as the one underlying the doped layer should give an easy access to the junction depth of the sample if the pump and probe beams are sufficiently spaced apart. In other words, the junction depth of a doped layer can be determined by

$$X_j = \frac{\lambda_{probe}}{4\pi n_0} \arccos\left[\lim_{r\to\infty} \frac{\Delta R_{ac}^{Box}(r)}{\Delta R_{ac}^{Substrate}(r)}\right]$$
$$\text{if } X_j \le \frac{\lambda_{probe}}{4n_0} \approx 44\,\text{nm} \tag{7.4a}$$

$$X_j = \frac{\lambda_{probe}}{2n_0} - \frac{\lambda_{probe}}{4\pi n_0} \arccos\left[\lim_{r\to\infty} \frac{\Delta R_{ac}^{Box}(r)}{\Delta R_{ac}^{Substrate}(r)}\right]$$
$$\text{if } \frac{\lambda_{probe}}{4n_0} \approx 44\,\text{nm} < X_j \le \frac{\lambda_{probe}}{2n_0} \approx 88\,\text{nm}, \tag{7.4b}$$

if we limit our study to sub-88 nm layers. These are our equations for the *absolute* determination of junction depths. Notice that, Eqs. (7.4a) and (7.4b) show that the derived X_j is unique only if one focuses on sub-44 nm junction depths. For real junction depths of USJs, this is obviously not a problem ($X_j < 20$ nm). In our investigations, however, both formulas (7.4a) and (7.4b) are needed (CVD12 and CVD13 have $X_j \le 60$ nm).

Before testing this technique on real measurements, we would like to point out that the right-hand side of Eq. (7.3) is a real constant. As a consequence, concerning the left-hand side, it is expected that:

- its amplitude should vary in the first one or two micrometer separation, corresponding to the faster decay in thermal and layer plasma components. For larger separations, its amplitude should saturate and become independent from the beam separation,
- its saturation value at large beam separation should correlate with $\cos(4\pi n_0 X_j/\lambda_{probe})$, i.e. its amplitude should be included between 0 and 1,
- its phase can have any value in the first one or two micrometers but should be equal to 0° or ±180° for larger separations.

Any other behavior would be in conflict with Eq. (7.3). Interestingly, this provides a verification procedure for the safe use of the developed X_j determination technique.

Let us apply this technique to the CVD12 matrix, the offset curves of which are shown in Fig. 6.3. Based on these experimental curves, we calculate the ratio $\Delta R_{ac}^{Box}/\Delta R_{ac}^{Substrate}$, the amplitude and phase of which are shown in Fig. 7.1a and b respectively.

Let us first look at the behavior of the amplitude of the ratio shown in Fig. 7.1a. It can be observed that, as expected from Eq. (7.3), the curves corresponding to

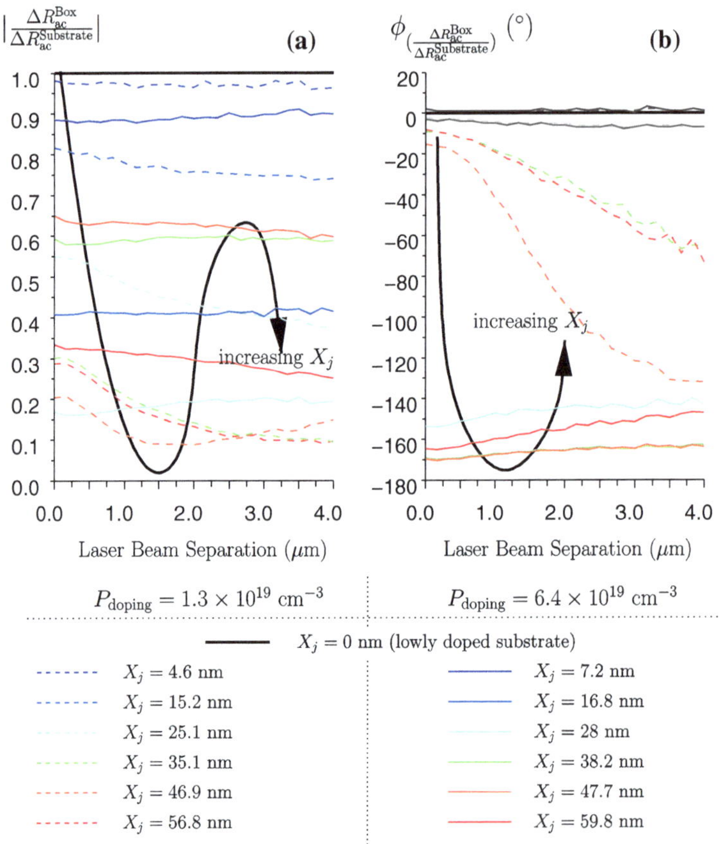

Fig. 7.1 Offset *curves*, i.e. variation with laser beam separation, of **a** the amplitude and **b** the phase of the ratio of the AC reflectance ΔR_{ac}^{Box} measured on p-type layers divided by the AC reflectance $\Delta R_{ac}^{Substrate}$ measured on a lowly doped substrate (*left-hand* side of Eq. (7.3)), as measured on the CVD12 matrix. The junction depths range from 0 to 60 nm and the doping concentrations are 1.3×10^{19} cm^{-3} (*dashed lines*) and 6.4×10^{19} cm^{-3} (*full lines*). These active doping concentration values are based on sheet resistance measurements and an assumed bulk mobility. All layers were grown on lowly doped device-quality substrates, the behavior of which is shown by the *thick black line*. The *arrows* show the trend observed when increasing junction depth

shallow junctions ($\leq$35 nm) flatten after one or two micrometer separation. Further, the saturation value of all curves lies between 0 and 1. However, for deeper junctions, the expected saturation is not observed, invalidating Eq. (7.3). Note that the saturation behavior of all curves can be explained by a decaying layer plasma component (see discussion of the accuracy in Sect. 7.1.1.2). In conclusion, from the behavior of the amplitude of the ratio, it seems that the technique could only be applied successfully to sub-35 nm junction depths.

The behavior of the phase of the ratio shown in Fig. 7.1b confirms the observations made on the amplitude. The phase of the ratio is indeed $\approx$0 for ultra-shallow layers, independently from the doping concentration in the layer. However, for deeper layers, we observe that the phase goes down to $-180°$ but never reaches that value. This behavior is also a consequence of the presence of a non negligible layer plasma component.

In other words, the verification procedure is only validated in a few cases, which tends to indicate that the technique may lead to erroneous junction depth values. To confirm this statement, we compare the TP *absolute* junction depths, as derived from this technique, to the values measured with SIMS on the CVD12 and CVD13 matrices. We, in particular, use the value of the ratio at 4 μm separation to obtain the results shown in Fig. 7.2. The correlation is acceptable for the ultra-shallow samples ($X_j < 20$ nm), i.e. for the layers which complied with the verification procedure. For deeper layers, however, the correlation is quite poor. In the next sections, we further discuss the correlation observed in Fig. 7.2 and determine whether the lack of correlation is due to a low precision (Sect. 7.1.1.1) or a low accuracy (Sect. 7.1.1.2).

Let us make two final remarks about the results shown in Fig. 7.2. First, as already pointed out by Eq. (7.4), the technique cannot distinguish between a junction depth X_j and $[\lambda_{\text{probe}}/(2n_0) - Xj]$. In Fig. 7.2, we have therefore used the known SIMS junction depth to decide which formula to apply (see the two different zones separated by the vertical dotted line). Second, in order to derive a value of the junction depth, we have assigned a positive sign to the amplitude of the ratio when the phase of the ratio is included between 0 and $-90°$ and a negative sign when it is included between -90 and $-180°$. Only then can a value of the TP absolute junction depth be derived from formulas (7.4). Note that this questionable definition of the sign is only needed because the experimental data do not comply with the safety check. Equation (7.3) indeed predicts that the ratio $\Delta R_{\text{ac}}^{\text{Box}}/\Delta R_{\text{ac}}^{\text{Substrate}}$ should be real so that the sign definition should be automatic.

7.1.1.1 Precision

Based on the signal repeatability $\sigma_{\Delta R_{\text{ac}}} = 7 \times 10^{-4}$ (a.u.) calculated in Appendix A.2, let us determine the precision σ_{X_j} of the technique. This requires the following derivation

$$\begin{aligned}
\sigma_{X_j} &= \left| \frac{\partial X_j}{\partial \frac{\Delta R_{\text{ac}}^{\text{Box}}}{\Delta R_{\text{ac}}^{\text{Substrate}}}} \right| \sigma_{\left(\frac{\Delta R_{\text{ac}}^{\text{Box}}}{\Delta R_{\text{ac}}^{\text{Substrate}}}\right)} \\
&= \frac{\lambda_{\text{probe}}}{4\pi n_0 \left|\sin(4\pi n_0 X_j/\lambda_{\text{probe}})\right|} \\
&\quad \times \left[\left| \frac{\partial}{\partial \Delta R_{\text{ac}}^{\text{Box}}} \frac{\Delta R_{\text{ac}}^{\text{Box}}}{\Delta R_{\text{ac}}^{\text{Substrate}}} \right| \sigma_{\Delta R_{\text{ac}}} + \left| \frac{\partial}{\partial \Delta R_{\text{ac}}^{\text{Substrate}}} \frac{\Delta R_{\text{ac}}^{\text{Box}}}{\Delta R_{\text{ac}}^{\text{Substrate}}} \right| \sigma_{\Delta R_{\text{ac}}} \right]
\end{aligned}$$

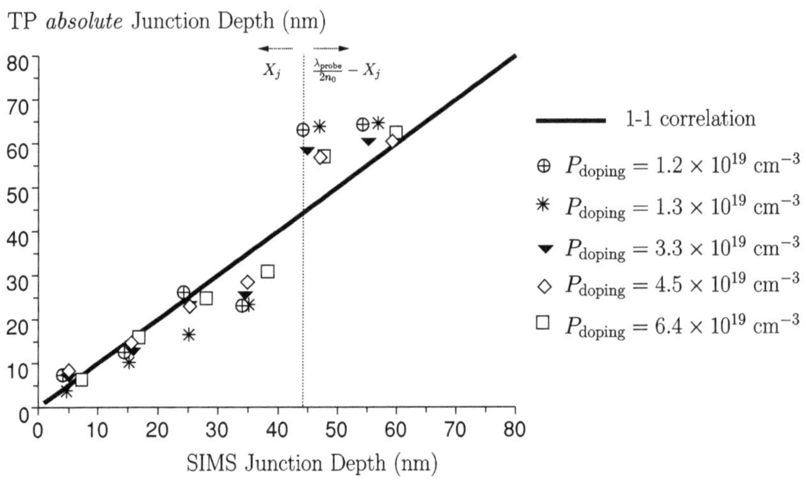

Fig. 7.2 Correlation between the TP *absolute* junction depths derived from Eq. (7.4) and the SIMS junction depths for the CVD12 and CVD13 matrices, i.e. five sets of B-doped CVD box-like profiles with active doping concentrations ranging between 1.2×10^{19} and 6.4×10^{19} cm^{-3} (active doping concentration values based on sheet resistance measurements and an assumed bulk mobility). The correlation is acceptable for ultra-shallow junctions but poor for deeper layers. The *thick black line* corresponds to 100 % correlation. The *vertical dotted line* indicates the separation for the application of Eq. (7.4a) (*left*) or (7.4b) (*right*)

$$= \frac{\lambda_{\text{probe}}}{4\pi n_0 \left|\sin(4\pi n_0 X_j/\lambda_{\text{probe}})\right|}(1 + |\cos(4\pi n_0 X_j/\lambda_{\text{probe}})|)\frac{\sigma_{\Delta R_{\text{ac}}}}{|\Delta R_{\text{ac}}^{\text{Substrate}}|}, \tag{7.5}$$

where $\sigma_{(\Delta R_{\text{ac}}^{\text{Box}}/\Delta R_{\text{ac}}^{\text{Substrate}})}$ is the cumulated precision of the subscripted ratio. The resulting precision of the determined junction depth is shown in Fig. 7.3 as a function of junction depth. Sub-nanometer junction depth precision is observed for almost all junction depths, except around 0 and 44 nm, i.e. around the extrema of Eq. (7.3). In other words, it can be concluded that the poor correlation with SIMS is not due to random errors.

It is important to note that, as a consequence of the presence of the substrate signal in Eq. (7.5), the precision drops with beam separation as can be observed from the comparison of the curves respectively obtained with coincident beams and with a 4 μm beam separation. Nevertheless, the precision remains very satisfactory even with a 4μm beam separation.

7.1.1.2 Accuracy

Since the random error proves to be low, it can be expected that the poor correlation of Fig. 7.2 is explained by systematic errors, i.e. a low accuracy. As a verification

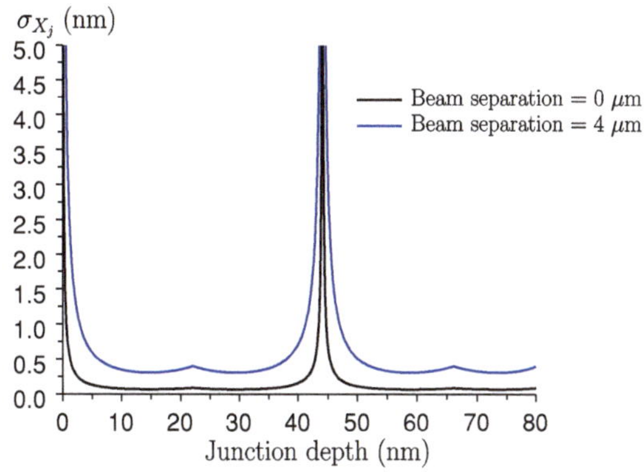

Fig. 7.3 Variation as a function of the junction depth in the precision σ_{X_j} of the TP absolute junction depths derived from Eq. (7.4) when using aligned beams (*black curve*) and a 4 μm beam separation (*blue curve*). For aligned beams, a sub-nm precision is achieved for all depths included between 1 and 43 nm and between 45 and 87 nm. The precision drops with larger beam separation

of this statement, the residual $[\Delta R_{ac}^{Box}(r)/\Delta R_{ac}^{Substrate}(r) - \cos(4\pi n_0 X_j/\lambda_{probe})]$ of formula (7.3) is shown for all the layers of matrices CVD12 and CVD13 in Fig. 7.4, where it can be observed that the residual follows a cosinusoidal behavior. Further, the amplitude of the observed cosine decreases with increasing doping in the layer. The corresponding phase, however, proves to be close to zero. Note that the noise observed on the phase for sub-10 nm layers is due to the low magnitude of the residual (phase is indefinite if residual is zero) and should not be understood as a large error.

From these observations, it is actually fairly simple to prove that the systematic error is due to the presence of a non negligible layer plasma component. Indeed, taking the ratio of Eqs. (6.10) and (6.8) and considering the layer plasma and thermal components as second-order effects, we obtain

$$\frac{\Delta R_{ac}^{Box}(r)}{\Delta R_{ac}^{Substrate}(r)} - \cos(4\pi n_0 X_j/\lambda_{probe}) \approx \frac{\Delta N_{l1}(r) - \frac{\delta}{\beta(1/m_e+1/m_h)}\Delta T_1(r)}{\Delta N_{sub1}(r)}[1 - \cos(4\pi n_0 X_j/\lambda_{probe})], \quad (7.6)$$

where we immediately recognize the cosine behavior followed by the amplitude of the residual shown in Fig. 7.4a and b. Note, furthermore, that the layer plasma and thermal components can be discriminated from the behavior of the phase of the residual in Fig. 7.4c and d, which indicate a dominant layer plasma component (phase $\approx$0). The phase of the residual would be around 180° if the thermal component were dominant.

In summary, the low correlation with SIMS is due to a limited accuracy induced by the non negligible presence of the layer plasma component even at a 4 μm beam separation. Note that the comparison of Fig. 7.4a and b confirms that the decay length of the layer plasma and thermal components is shorter than that of the substrate plasma component. However, according to the observed decay in $\Delta R_{ac}^{Box}/\Delta R_{ac}^{Substrate}$, approximate calculations show that a laser beam separation of 20–30 μm would be

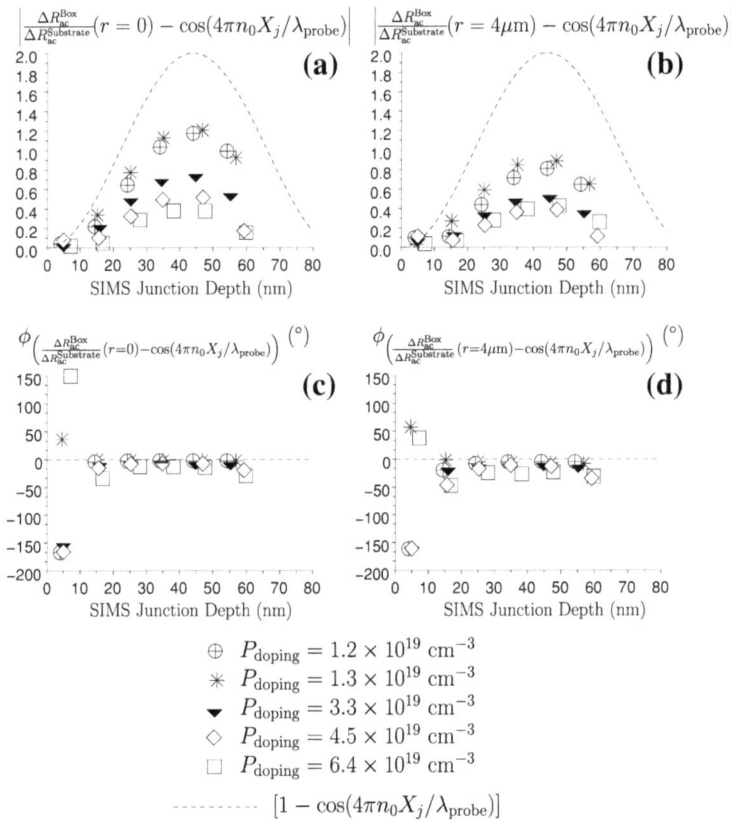

Fig. 7.4 Variation as a function of the SIMS junction depth in the (*top*) amplitude and (*bottom*) phase of the residual $[\Delta R_{ac}^{Box}(r)/\Delta R_{ac}^{Substrate}(r) - \cos(4\pi n_0 X_j/\lambda_{probe})]$ of Eq. (7.3) in case of (*left*) coincident beams and (*right*) a 4 μm beam separation. The behavior of the residual is due to a non-negligible layer plasma component, even at a 4 μm beam separation. This leads to a large systematic error. The *black interrupted line* indicates the behavior of the function $[1 - \cos(4\pi n_0 X_j/\lambda_{probe})]$

needed for $\Delta N_{l1}/\Delta N_{sub1}$ to be less than 10 %. At these separations, however, the substrate injection is so low that the technique precision would be unacceptably reduced, even if higher pump irradiances were used. We, therefore, believe that, unless a method is found to reduce the decay length of the layer plasma component, this absolute junction depth determination technique is bound to remain inaccurate.

Note that the technique could have been promising when applied on highly activated implanted layers (lower amplitude and shorter decay length of the layer plasma component) but unfortunately it cannot be extended to annealed implanted layers due to the impossible isolation of the substrate plasma component (integral convolution) in Eq. (6.17c).

As a final remark, note also that a similar technique can easily be developed to measure *relative* junction depths, as we do in the next section. The difference with the theory presented here is that the reference measurement (denominator of Eq. (7.3)) is measured on another doped layer of known junction depth [1]. We, however, do not discuss this other technique since the conclusions reached would be very similar to the ones of the present section. In particular, low accuracy is achieved with the current experimental setup of the TP tool.

To conclude, offset curves can, in principle, be used to determine the junction depth of box-like profiles with a sub-nm precision almost independently from the junction depth. The accuracy of the technique, as applied with the TP630XP tool, however, is poor. An improvement should result from a larger beam separation, at the expense of a reduced precision. The combination of a high precision and a high accuracy would require a more efficient way to reduce the layer plasma component.

7.1.2 Relative Determination of Junction Depths

Based on the conclusions of the previous section, it appears that one needs to find another approach to get rid of the thermal and layer plasma components. In this section, we propose an alternative technique which, though less flexible, proves to be in excellent agreement with SIMS. Further, all measured layers are compliant with the verification procedure which this technique also offers. This, very interestingly, allows us to check some of the assumptions we have made so far.

In order to cancel the thermal component, rather than dividing $\Delta R_{\text{ac}}^{\text{Box}}$ by $\Delta R_{\text{ac}}^{\text{Substrate}}$ like in Sect. 7.1.1 above, we here propose to subtract them. Given the layer-independence of the fundamental modes of the excess carrier and temperature distributions in the substrate, this gives

$$\Delta R_{\text{ac}}^{\text{Box}}(r) - \Delta R_{\text{ac}}^{\text{Substrate}}(r) = \frac{4R_0}{n_0^2 - 1}\Gamma_0 \exp(-i\theta_0)\beta\left(\frac{1}{m_e} + \frac{1}{m_h}\right) \times [\Delta N_{sub1}(r) - \Delta N_{l1}(r)][1 - \cos(4\pi n_0 X_j/\lambda_{\text{probe}})], \quad (7.7)$$

where, importantly, the prefactors of the substrate and layer excess carrier concentrations are now the same. An extra measurement $\Delta R_{\text{ac}}^{\text{Ref}}(r)$ on another doped layer, with the same active doping and with a known junction depth X_j^{Ref}, suffices to cancel the layer plasma component. In particular, we apply Eq. (7.7) respectively on the unknown layer and on the reference layer and calculate their ratio, which yields

$$\frac{\Delta R_{\text{ac}}^{\text{Box}}(r) - \Delta R_{\text{ac}}^{\text{Substrate}}(r)}{\Delta R_{\text{ac}}^{\text{Ref}}(r) - \Delta R_{\text{ac}}^{\text{Substrate}}(r)} = \frac{1 - \cos(4\pi n_0 X_j/\lambda_{\text{probe}})}{1 - \cos(4\pi n_0 X_j^{\text{Ref}}/\lambda_{\text{probe}})}, \quad (7.8)$$

where the only remaining unknown is the junction depth X_j. Limiting the study to sub-88 nm junction depths, we have

$$X_j = \frac{\lambda_{\text{probe}}}{4\pi n_0} \arccos\left[1 - \left(1 - \cos(4\pi n_0 X_j^{\text{Ref}}/\lambda_{\text{probe}})\right) \frac{\Delta R_{\text{ac}}^{\text{Box}}(r) - \Delta R_{\text{ac}}^{\text{Substrate}}(r)}{\Delta R_{\text{ac}}^{\text{Ref}}(r) - \Delta R_{\text{ac}}^{\text{Substrate}}(r)}\right]$$

$$\text{if } X_j \leq \frac{\lambda_{\text{probe}}}{4n_0} \approx 44\,\text{nm} \tag{7.9a}$$

$$X_j = \frac{\lambda_{\text{probe}}}{2n_0} - \frac{\lambda_{\text{probe}}}{4\pi n_0} \arccos\left[1 - \left(1 - \cos(4\pi n_0 X_j^{\text{Ref}}/\lambda_{\text{probe}})\right) \frac{\Delta R_{\text{ac}}^{\text{Box}}(r) - \Delta R_{\text{ac}}^{\text{Substrate}}(r)}{\Delta R_{\text{ac}}^{\text{Ref}}(r) - \Delta R_{\text{ac}}^{\text{Substrate}}(r)}\right]$$

$$\text{if } \frac{\lambda_{\text{probe}}}{4n_0} \approx 44\,\text{nm} < X_j \leq \frac{\lambda_{\text{probe}}}{2n_0} \approx 88\,\text{nm}. \tag{7.9b}$$

We call the values obtained using Eqs. (7.9a) and (7.9b) the TP *relative* junction depths.

A verification procedure can be derived from Eqs. (7.7) and (7.8). In particular, the following experimental behavior is expected:

- the amplitude of the ratio calculated from the left-hand side of Eq. (7.8) should be independent from the laser beam separation,
- the phase of the difference calculated from the left-hand side of Eq. (7.7) should be independent from the layer since only plasma components contribute to this expression. As a consequence, the phase of the ratio calculated from the left-hand side of Eq. (7.8) should be equal to zero, independently from the laser beam separation.

Notice that, contrary to the technique developed in the previous section, the absolute value of the ratio can have any value, according to the reference junction depth X_j^{Ref}.

Let us apply this technique to the layers of the CVD12 matrix, the offset curves of which are shown in Fig. 6.3. A different reference sample is needed for each doping concentration. The CVD12 matrix is composed of two sets of layers with two different doping concentrations. For each, we choose the samples with a SIMS junction depth close to 44 nm and justify our choice in Sects. 7.1.2.1 and 7.1.2.2 below. The application of Eqs. (7.7) and (7.8) gives the behavior shown in Fig. 7.5.

Let us first discuss the behavior of the amplitude of the differences and ratios, respectively shown in Fig. 7.5a and c. First, as expected, the amplitude of the ratio is independent from laser beam separation on almost all measured samples. Second, since $X_j^{\text{Ref}} \approx 44$ nm, i.e. the reference junction depths are very close to $\lambda_{\text{probe}}/(4n_0)$, it is easy to recognize the non-monotonic cosine behavior. In this specific case, the amplitude of the ratio should indeed be included between 0 and 1.

Concerning the phase of the differences and ratios, shown respectively in Fig. 7.5b and d, their behavior is also in excellent agreement with our verification procedure. First, the phase of the difference is the same for almost all layers. Besides, the plasma wavelength is easily recognized in the lateral behavior ($\Lambda_1^{pl} \approx 100\,\mu$m, Fig. 4.14f).

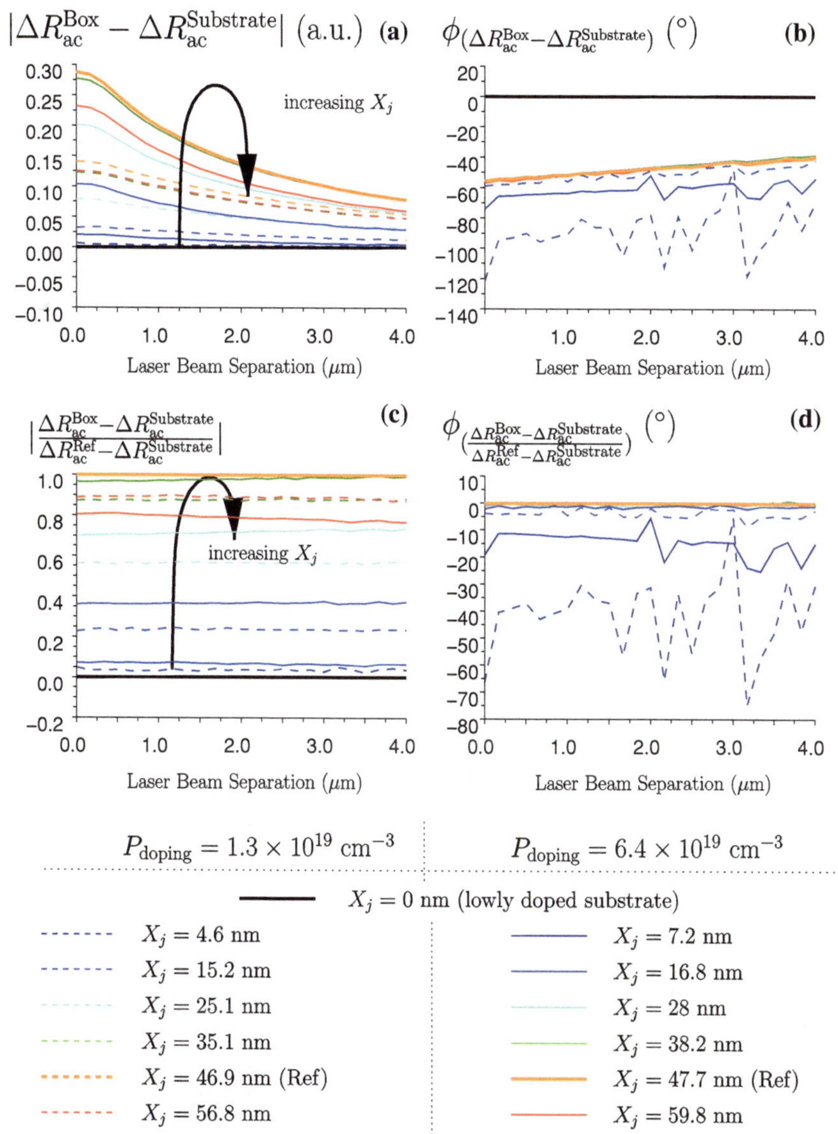

Fig. 7.5 Offset *curves*, i.e. variation with laser beam separation, of (*left*) the amplitude and (*right*) the phase of (*top*) the *left-hand* side of Eq. (7.7) and (*bottom*) the *left-hand* side of Eq. (7.8), as measured on the CVD12 matrix. The samples with a SIMS junction depth close to 44 nm have been used as references (*thick orange lines*). The junction depths range from 0 to 60 nm and the doping concentrations are 1.3×10^{19} cm^{-3} (*dashed lines*) and 6.4×10^{19} cm^{-3} (*full lines*). These active doping concentration values are based on sheet resistance measurements and an assumed bulk mobility. All layers were grown on lowly doped device-quality substrates, the behavior of which is shown by the *thick black line*. The *arrows* show the trend observed when increasing junction depth

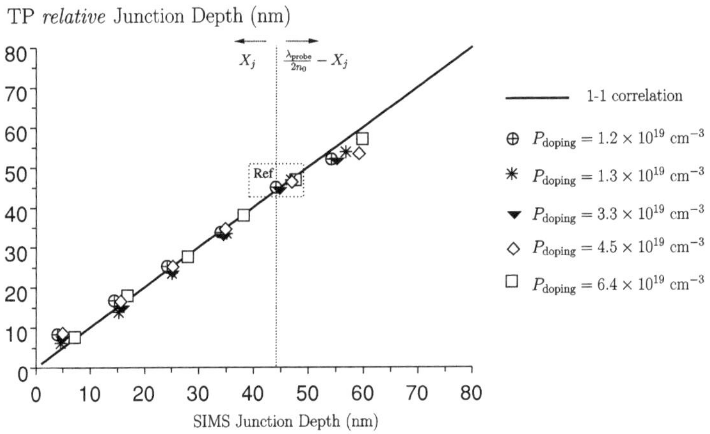

Fig. 7.6 Correlation between the TP *relative* junction depths derived from Eqs. (7.9a) and (7.9b) and the SIMS junction depths for the CVD12 and CVD13 matrices, i.e. five sets of B-doped CVD box-like profiles with active doping concentrations ranging between 1.2×10^{19} and 6.4×10^{19} cm^{-3} (active doping concentration values based on sheet resistance measurements and an assumed bulk mobility). For each doping concentration, the reference sample has $X_j^{Ref} \approx 44$ nm, as highlighted by the *dotted box*. The correlation is very good though a small deviation remains, especially noticeable on sub-20 nm junctions. The *thick black line* corresponds to 100 % correlation. The *vertical dotted line* indicates the separate regions where respectively Eq. (7.9a) (*left*) and Eq. (7.9b) (*right*) are used

Note that, owing to the normalization phase θ_0, the phase for coincident beams is about $-60°$, which is different from the 7° phase of the plasma wave calculated in Fig. 4.14e ($\theta_0 \approx -60° + 7° \approx -53°$). Second, the phase of the ratio is equal to zero independently from laser beam separation on almost all the layers. The noise and error observed on shallow layers is simply due to the fact that the phase of a complex number with an absolute value close to zero is indefinite.

The perfect agreement between these experimental observations and our theoretical expectations shows that the technique is very likely to work. This is confirmed by Fig. 7.6 showing the excellent correlation between the TP relative junction depths, as obtained from Eqs. (7.9a) and (7.9b) using the data for coincident beams, and the SIMS junction depths of all the layers of the CVD12 and CVD13 matrices. The quality of the observed correlation is further discussed in Sects. 7.1.2.1 and 7.1.2.2 where we respectively look at the precision and accuracy of the technique.

As an important final remark, we would like to point out that the agreement between all these experimental observations and our theory gives us the opportunity to confirm some of the assumptions underlying our theory. First, the behavior observed in Fig. 7.5b shows that the fundamental mode of the excess carrier concentrations ΔN_{sub_1} and ΔN_{l1} respectively in the substrate and in the doped layer have the same phase and wavelength. Further, it also shows that the fundamental mode ΔT_1 of the temperature distribution is fully independent from the doped layer. Second, the flat offset curves in Fig. 7.5c show that the position of the probe beam during an offset curve measurement has a negligible impact on the value of ΔR_{ac}. It also confirms the infinite abruptness of the interface (see Sect. 7.2.1 for dependence of offset curves upon abruptness). Third, the independence from the doped layer of the fundamental mode ΔN_{sub1} of the excess carrier in the substrate is a necessary condition for the behavior observed in Fig. 7.5b and c. Finally, as discussed in Sect. 7.1.2.2 below, the excellent correlation observed in Fig. 7.6 is also a confirmation of the negligible magnitude of the electroabsorptive BGN effect.

7.1.2.1 Precision

Based on the signal repeatability $\sigma_{\Delta R_{\text{ac}}} = 7 \times 10^{-4}$ (a.u.) determined in Appendix A.2, we would like to calculate the precision σ_{X_j} of the technique. This, in particular, allows us to determine whether random error can explain the small deviations observed in the correlation plot of Fig. 7.6. The precision can be calculated as follows

$$\sigma_{X_j} = \left| \frac{\partial X_j}{\partial \frac{\Delta R_{\text{ac}}^{\text{Box}} - \Delta R_{\text{ac}}^{\text{Substrate}}}{\Delta R_{\text{ac}}^{\text{Ref}} - \Delta R_{\text{ac}}^{\text{Substrate}}}} \right| \sigma \left(\frac{\Delta R_{\text{ac}}^{\text{Box}} - \Delta R_{\text{ac}}^{\text{Substrate}}}{\Delta R_{\text{ac}}^{\text{Ref}} - \Delta R_{\text{ac}}^{\text{Substrate}}} \right)$$
$$= \frac{\lambda_{\text{probe}}}{4\pi n_0} \frac{1}{\left|\sin(4\pi n_0 X_j/\lambda_{\text{probe}})\right| \left|1 - \Delta N_{l1}/\Delta N_{sub1}\right|} \frac{\sigma_{\Delta R_{\text{ac}}}}{\left|\Delta R_{\text{ac}}^{\text{Substrate}}\right|}$$
$$\times \left[1 + \frac{\left|\cos(4\pi n_0 X_j^{\text{Ref}}/\lambda_{\text{probe}}) - \cos(4\pi n_0 X_j/\lambda_{\text{probe}})\right|}{[1 - \cos(4\pi n_0 X_j^{\text{Ref}}/\lambda_{\text{probe}})] \left|1 - \Delta N_{l1}/\Delta N_{sub1}\right|} + \frac{1 - \cos(4\pi n_0 X_j/\lambda_{\text{probe}})}{1 - \cos(4\pi n_0 X_j^{\text{Ref}}/\lambda_{\text{probe}})} \right], \quad (7.10)$$

where $\sigma_{[(\Delta R_{\text{ac}}^{\text{Box}} - \Delta R_{\text{ac}}^{\text{Substrate}})/(\Delta R_{\text{ac}}^{\text{Ref}} - \Delta R_{\text{ac}}^{\text{Substrate}})]}$ is the precision of the ratio in subscript. The resulting technique precision is shown as a function of junction depth in Fig. 7.7 for three different values of the reference junction depth X_j^{Ref} (15, 30 and 44 nm) and two different values of the injection ratio $\Delta N_{l1}/\Delta N_{sub1}$ (0 and 0.5 corresponding respectively to ultra high doping and moderate doping). In the case studied here, i.e. $X_j^{\text{Ref}} \approx 44$ nm and $\Delta N_{l1}/\Delta N_{sub1} \approx 0.5$, the precision is sub-nanometric for junction depths included between 5 and 39 nm as well as between 49 and 84 nm.

Figure 7.7 also underlines the importance of a well chosen reference layer. The ideal X_j^{Ref} is close to 44 nm, which justifies the choice we have made in this study. A change in this value can be very detrimental to the technique precision.

In conclusion, though some precision has been lost compared to the technique studied in Sect. 7.1.1, the small deviations observed in the correlation of Fig. 7.6 cannot be explained by random errors.

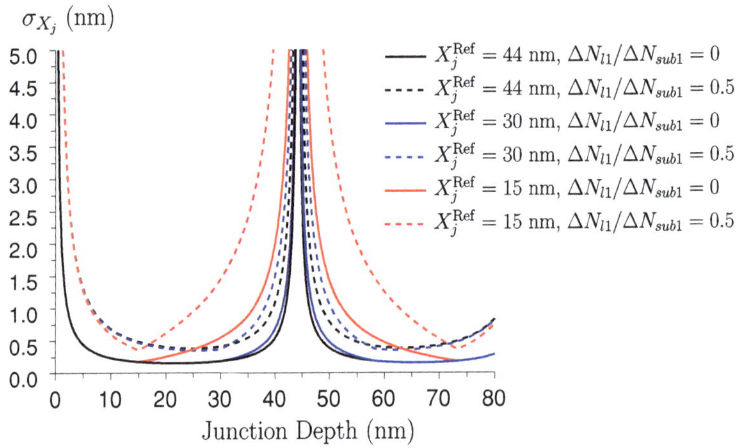

Fig. 7.7 Variation as a function of the junction depth in the precision σ_{X_j} of the TP relative junction depth determination technique for three different values of X_j^{Ref} and two different values of the injection ratio $\Delta N_{l1}/\Delta N_{sub1}$. Aligned beams are assumed. For the case studied here ($X_j^{Ref} \approx 44$ nm and $\Delta N_{l1}/\Delta N_{sub1} \approx 0.5$), a sub-nm precision is achieved for all depths included between 5 and 39 nm and between 49 and 84 nm

7.1.2.2 Accuracy

To verify whether systematic errors on the derived TP relative junction depths are the cause for the small deviations observed in Fig. 7.6, we need to identify the different possible sources of error. First, modeling errors could induce the deviation. Second, the propagation of a possible error in the characterization of the reference sample could also explain the deviation.

First, let us start the discussion by showing that no modeling error is expected to affect the developed technique. The TP relative junction depths have been derived using Eqs. (7.9a) and (7.9b), which only involve the dependence of the reflectance perturbation on the junction depth. This dependence has been initially derived in Chap. 2 and is based on three assumptions. First, the presence of a native oxide at the interface between silicon and air has been neglected. Section 2.5.1 indeed shows that the impact of the presence of a native oxide is negligible. Second, Sect. 2.5.2 shows that any abrupt lateral variation in reflectance perturbation could modify the integrated perturbed reflectance as measured by a laser with finite radius. However, as a direct consequence of the division in Eq. (7.8), the impact of the integration vanishes in this equation. Finally, the BGN electrooptical effect has been neglected. Yet, its consideration would induce a small shift in the TP relative junction depth. However, Sect. 3.2.2 has shown that this effect is not expected to be significant. Besides, as shown in Appendix A.6, it would systematically shift the TP relative junction depth towards deeper values. In summary, we do not believe that the observed deviation is due to a modeling error.

Second, it can be expected that the accuracy depends strongly on the reference sample. In this section, we investigate how sensitive the technique is to any error on the values of the junction depth or the active doping concentration of the reference sample. Interesting recommendations for the reference sample can be derived from these results.

Let us first discuss the sensitivity of the technique to an error on the junction depth X_j^{Ref} of the reference sample. Let us suppose an error $\epsilon_{X_j^{\text{ref}}}$ is made on this value. As a result of the error propagation, the error $\epsilon_{X_j}^{\text{depth}}$ made on the value of the TP relative junction depth is

$$\begin{aligned}\epsilon_{X_j}^{\text{depth}} &= \left|\frac{\partial X_j}{\partial X_j^{\text{ref}}}\right| \epsilon_{X_j^{\text{ref}}} \\ &= \frac{|\sin(4\pi n_0 X_j^{\text{ref}}/\lambda_{\text{probe}})|}{1-\cos(4\pi n_0 X_j^{\text{ref}}/\lambda_{\text{probe}})} \frac{1-\cos(4\pi n_0 X_j/\lambda_{\text{probe}})}{|\sin(4\pi n_0 X_j/\lambda_{\text{probe}})|} \epsilon_{X_j^{\text{ref}}}. \end{aligned} \tag{7.11}$$

The resulting error is plotted as a function of junction depth in Fig. 7.8 for three different values of X_j^{Ref} (15, 30 and 44 nm). It can be observed that, if $X_j^{\text{Ref}} = 44$ nm, the error made on the value of X_j^{Ref} has no impact whatsoever on the accuracy of the determine TP relative junction depth. An error made on the value of X_j^{Ref} can, however, be very problematic if another X_j^{Ref} is chosen. In other words, besides the high precision a 44-nm X_j^{Ref} offers, it also ensures a high accuracy. This is the second justification for choosing our reference samples in this depth range.

Let us now investigate the sensitivity of the technique to an error on the active doping concentration of the reference sample. Though the flat offset curves of Fig. 7.5c tend to show that this sensitivity is low, we believe it is worthwhile quantifying it. In particular, we want to calculate the error $\epsilon_{X_j}^{\text{doping}}$ due to an error $\epsilon_{P_{\text{doping}}^{\text{Ref}}}$ on the active doping concentration $P_{\text{doping}}^{\text{Ref}}$ of the reference sample, i.e.

$$\begin{aligned}\epsilon_{X_j}^{\text{doping}} &= \left|\frac{\partial X_j}{\partial P_{\text{doping}}^{\text{ref}}}\right| \epsilon_{P_{\text{doping}}^{\text{Ref}}} \\ &= \left|\frac{\partial X_j}{\partial \Delta N_{l1}^{\text{ref}}}\right| \left|\frac{\partial \Delta N_{l1}^{\text{ref}}}{\partial P_{\text{doping}}^{\text{ref}}}\right| \epsilon_{P_{\text{doping}}^{\text{Ref}}} \\ &= \frac{\lambda_{\text{probe}}}{4\pi n_0} \frac{1-\cos(4\pi n_0 X_j/\lambda_{\text{probe}})}{|\sin(4\pi n_0 X_j/\lambda_{\text{probe}})|} \frac{1}{|1-2\Delta N_{sub0}^{\text{Ref}}/(P_{\text{doping}}^{\text{Ref}})|} \\ &\quad \times \frac{2\Delta N_{sub0}^{\text{Ref}}}{P_{\text{doping}}^{\text{Ref}}} \frac{\epsilon_{P_{\text{doping}}^{\text{Ref}}}}{P_{\text{doping}}^{\text{Ref}}}, \end{aligned} \tag{7.12}$$

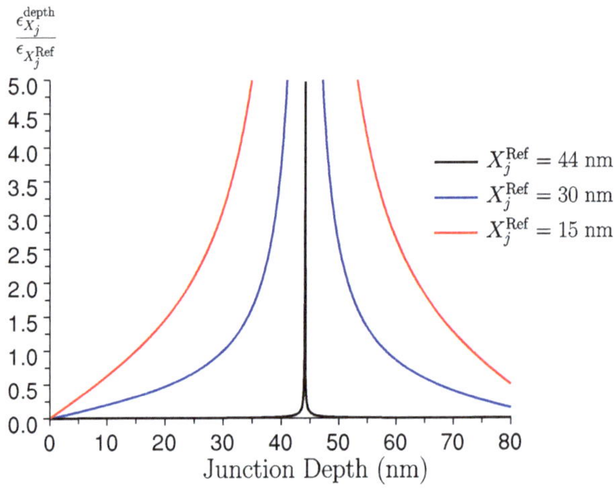

Fig. 7.8 Variation as a function of junction depth in the error $\epsilon_{X_j}^{\text{depth}}$ on the derived TP relative junction depth as a result of the propagation of the error $\epsilon_{X_j^{\text{ref}}}$ made on the junction depth of the reference sample. An error on the value of the junction depth of a 44 nm deep reference sample has no impact on the accuracy of the technique, except around 44 nm. The error, however, propagates significantly if another reference junction depth is used

where $\Delta N_{sub0}^{\text{Ref}}$, $\Delta N_{sub1}^{\text{Ref}}$ and $\Delta N_{l1}^{\text{Ref}}$ are respectively the constant and fundamental modes of the excess carrier distributions in the substrate of the reference sample and the fundamental mode of the excess carrier distribution in the reference layer. For the sake of simplicity of the final expression of Eq. (7.12), we have used Eq. (5.10) and neglected the BGN effect. The presence of the ratio $\Delta N_{sub0}^{\text{Ref}}/(P_{\text{doping}}^{\text{Ref}})$ in Eq. (7.12) explains the very low sensitivity of the technique to the active doping concentration of the reference sample (not shown). In other words, though the reference sample should be taken with a doping concentration as close as possible to that of the unknown sample, Eq. (7.12) shows that deviations, even large, do not impact the accuracy significantly.

As a final remark, note that the propagation of the error on the junction depth and doping concentration of the reference sample have here been discussed separately. The total error due to an erroneous characterization of the reference sample can easily be calculated from $\sqrt{(\epsilon_{X_j}^{\text{depth}})^2 + (\epsilon_{X_j}^{\text{doping}})^2}$.

In summary, we have shown that the technique for the determination of the TP relative junction depth has both high precision and high accuracy. In particular, the small disagreement with SIMS of Fig. 7.6 cannot be explained either by a random error or by a systematic error of the developed technique. Note that an alternative explanation lies in the systematic error made in the derivation of the SIMS junction depths (Sect. 1.2.1.1). First, the presence of an oxide at the surface of the samples leads SIMS to underestimating the junction depth, which could partially explain the deviation (systematic error). This would also explain why the observed discrepancy decreases with depth since this systematic error is expected to drop with increasing depth. Second, SIMS always underestimates the slope of profiles, which makes the definition of the SIMS junction depth difficult in the case of a box-like profile. In this work, the SIMS junction depths of CVD box-like profiles have been defined as the depths at which the SIMS profiles reach a doping concentration equal to 90 % of the peak doping concentration. We believe that, in spite of the systematic use of this definition, it is bound to induce some error as well.

7.1.2.3 Application

Given the good correlation with SIMS, the high precision and accuracy of the technique, we believe this technique shows great promise for the non-destructive measurement of relative junction depths e.g. over a whole wafer. As an illustration, we qualitatively compare in Fig. 7.9 a junction-depth map, as obtained with the introduced technique, with the sheet-resistance map measured with RsL on the same wafer (CVD12 2.2). We use the central point of the wafer as our needed reference measurement and deduce the variation in junction depth over the wafer. Figure 7.9a and b show respectively the obtained 49-point map of the junction depth and the 973-point maps of the sheet resistance. The shallow regions measured by TP correspond very clearly to the highly resistive regions measured by RsL, which indicates a good agreement between both techniques. Note that all 49 points of the junction depth map comply with the verification procedure.

Though promising, this type of application where the sought X_j is very close to X_j^{Ref} has to be considered with care. First, as shown in Fig. 7.7, sub-nanometer precision can be expected in most cases but junction depths close to 44 nm should be avoided. Further, as noted in Fig. 7.8, the accuracy of the external measurement of the reference junction depth is critical. Typically, the error on the derived X_j remains equal to the original error on X_j^{Ref} provided the variations are not too large. To keep both a high precision and a high accuracy, the use of the technique is recommended on highly doped layers ($>10^{19}\,\mathrm{cm}^{-3}$) and junction depths included between 10 and 35 nm.

As a final remark, note that the technique has been tested without success on annealed implanted layers. The reason for the failure lies in the impossible isolation of the substrate plasma component in the model Eq. (6.17c), which is a critical requirement for the developed technique.

In summary, we have developed a technique able to determine the junction depth of a box-like profile with a sub-nanometer precision and accuracy. This method has a high applicability potential in the uniformity test of the junction over full wafers. In this specific case, layers with depth in the neighborhood of 44 nm should be avoided.

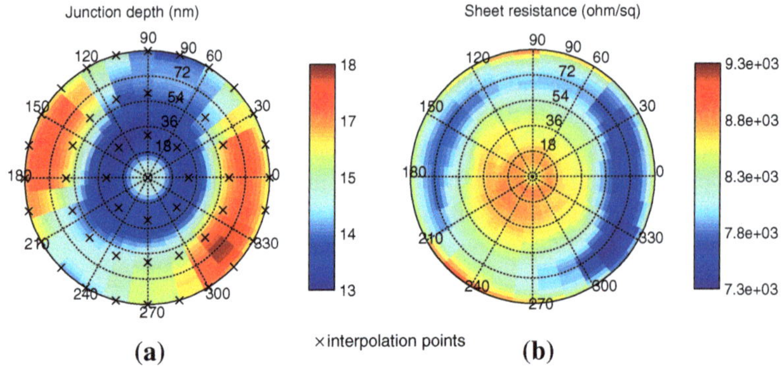

Fig. 7.9 Comparison of **a** the junction depth measured by TP and **b** the sheet resistance measured by RsL on a medium doped ($1.3\times10^{19}\,\text{cm}^{-3}$ in the *center*), shallow (15 nm in the *center*) CVD layer (CVD12 2.2). The color scale of each map is given on the right-hand side. The most resistive regions also present shallow TP depths, indicating a good agreement between TP and RsL. The TP junction depth map contains 49 measurement points (indicated by the *crosses*), while the sheet resistance map contains 973 measurement points (not shown). To facilitate the comparison, the two maps have been interpolated on the same grid. The junction depth map is obtained with coincident pump and probe beams and using the central point for the reference measurement

7.2 Model-Based Profile Characterization Technique

The discussion in this chapter has thus far been limited to the determination of junction depths. The ultimate objective of this work, however, is to evaluate whether, based on TP measurements, a method for the reconstruction of complete profiles can be developed. In other words, we are seeking a measurement technique offering information not only about the depth of profiles but also about their active doping concentration. Unfortunately, as it has been demonstrated, the sensitivity of ΔR_{ac} to doping tends to strongly decrease in heavily doped layers. As a consequence, ΔR_{ac}, by itself, is not a promising candidate for full profile reconstruction, especially considering the high active doping concentrations in modern USJs ($>10^{20}\,\text{cm}^{-3}$). Though this signal gives information about the depth and, as we show below, about the abruptness of active doping profiles, the doping information needs to be provided by another signal. On the other hand, R_{dc} has been theoretically and experimentally demonstrated to have a linear sensitivity to active doping concentration (Chap. 6). It therefore seems reasonable that the combination of ΔR_{ac} and R_{dc} should constitute a very promising approach for full profile reconstruction. Section 7.2.1 below gives more insight into this statement, based on a theoretical study of the profile information which can be retrieved from both signals. Section 7.2.2 then details how we have practically implemented these ideas into a technique. Finally, typical results are critically evaluated in Sect. 7.2.3. Note that this section is exploratory and therefore only reaches partial conclusions. We believe, however, that it gives a very bright outlook on the possible use of TP for carrier profiling.

7.2.1 Fundamental Principle of the Technique

Based on the theoretical understanding developed in this work, we want to show that the combined use of ΔR_{ac} and R_{dc} allows for the simultaneous measurement of the depth, abruptness and active doping concentration of a profile.

Let us start by investigating the dependencies of the AC reflectance. On a profile of arbitrary shape, ΔR_{ac} follows Eq. (6.17), which can be rewritten as

$$\Delta R_{ac}^{\text{Profile}}(r) = \frac{4R_0}{n_0^2 - 1}\Gamma_0 \exp(-i\theta_0) \times \Bigg\{ \underbrace{-\beta\left(\frac{1}{m_e} + \frac{1}{m_h}\right)\Delta N_{l1}(r, z = 0) + \delta\Delta T_1(r)}_{\text{surface}} \underbrace{- \beta\left(\frac{1}{m_e} + \frac{1}{m_h}\right)\int_{0+}^{+\infty} \frac{\partial \Delta N_{l1}(r, z)}{\partial z} \cos(4\pi n_0 z/\lambda_{\text{probe}})dz}_{\text{interface}} \Bigg\}. \tag{7.13}$$

In analogy with the case of box-like profiles, ΔR_{ac} can be understood as the coherent sum of two reflections. The first reflection occurs at the surface and is due to both the surface excess carriers $\Delta N_{l1}(z = 0)$ and the excess temperature ΔT_1. The second reflection is the interface reflection which, in this general case, is due to the multiple reflections occurring at each depth of the excess carrier profile (integral of Eq. (7.13)). Figure 7.10 illustrates this analogy. More particularly, Fig. 7.10a shows two active doping profiles with different slopes and Fig. 7.10b the subsequent excess carrier profiles for two different excess carrier concentrations in the substrate (10^{17} and 10^{18} cm^{-3}). The behavior of the integrand of Eq. (7.13) is shown in Fig. 7.10c for all four situations. On the one hand, in the case of a box-like profile, the integrand of Eq. (7.13) reduces to a peak at the junction,[2] i.e. the interface reflection is composed of one single large reflection. Furthermore, the position of this peak is independent from the substrate injection. On the other hand, when the slope of the profile is finite, the multiple reflections at the interface broaden the peak. The position of the maximum of this peak, i.e. the maximum interface reflection, can be understood as the depth ΔR_{ac} is sensitive to.

As can already be observed in Fig. 7.10c, in the case of a profile with a finite slope, the maximum interface reflection moves according to the substrate injection. This can be proven mathematically as follows. Neglecting the cosine envelope of the integrand, the position z_{max} of the maximum interface reflection can be defined as the position at which the excess carrier profile is the steepest, i.e. an approximate

[2] Note that the absolute height of the peak shown in Fig. 7.10 is irrelevant since it depends on the discretization of the depth scale. For completeness, the results shown in figure assume a 0.01 nm step size.

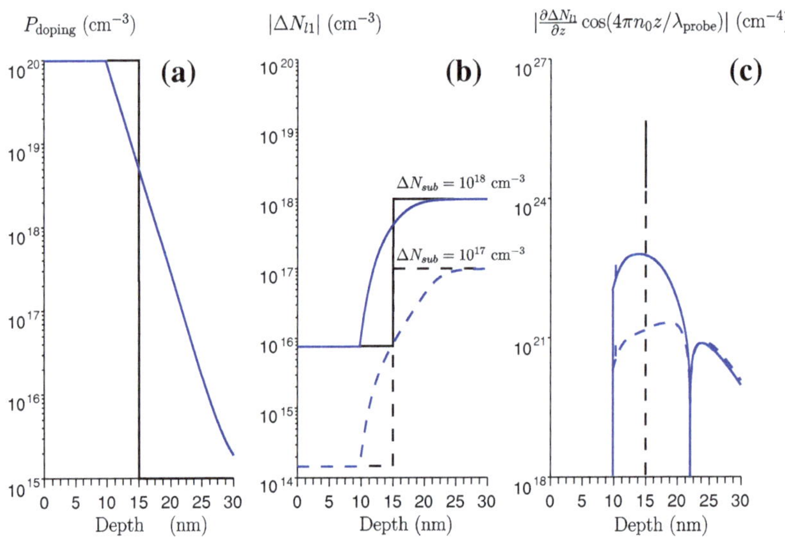

Fig. 7.10 **a** Comparison of a profile of infinite slope, i.e. a box-like profile (*black*) with a profile of finite slope (*blue*). **b** Subsequent excess carrier profiles when the substrate carrier concentration is of $10^{17}\,\mathrm{cm}^{-3}$ (*interrupted lines*) and $10^{18}\,\mathrm{cm}^{-3}$ (*full lines*). The colors correspond to the active doping profiles of (**a**). **c** Behavior of the integrand of Eq. (7.13) for the four excess carrier profiles of (**b**)

value of z_{max} can be found by solving the following implicit equation

$$\frac{\partial^2 \Delta N_{l1}(z)}{\partial z^2}\Big|_{z=z_{max}} = 0$$

$$\Leftrightarrow \frac{\dfrac{\partial^2[P_{doping}(z_{max})/\Delta N_{sub}]}{\partial z^2}}{\left\{\dfrac{\partial[P_{doping}(z_{max})/\Delta N_{sub}]}{\partial z}\right\}^2} = \frac{1}{\sqrt{4 + P_{doping}^2(z_{max})/\Delta N_{sub}^2}}\left(1 + \frac{1}{\sqrt{1 + 4\Delta N_{sub}^2/P_{doping}^2(z_{max})}}\right), \tag{7.14}$$

where we have made use of Eq. (5.8) assuming Bolzmann statistics and neglecting BGN. Though Eq. (7.14) may seem complex, it simply signifies that the depth of the maximum interface reflection depends solely on the function $P_{doping}(z)/\Delta N_{sub}$ and its derivatives. For a better understanding of the meaning of this equation, let us take the example of an active doping profile decaying with an exponential tail, as defined e.g. in Eq. (7.16) below. In this particular case, using Eq. (7.14), the maximum interface reflection occurs at a depth $z_{\max}$ such that

$$\frac{P_{\text{doping}}(z = z_{\max})}{\Delta N_{sub}} = \sqrt{2(-1 + \sqrt{5})} \approx 1.57, \tag{7.15}$$

i.e. the maximum interface reflection comes from the depth where the doping concentration is $1.57 \times \Delta N_{sub}$. From this result, it is correct to extrapolate that the maximum interface reflection always originates from a depth where the doping is commensurate with the substrate injection, independently from the profile itself. Note that the particular value obtained in Eq. (7.15), however, cannot be generalized to all profiles since the profile derivatives are involved into Eq. (7.14).

In conclusion, our theory shows that ΔR_{ac} is sensitive to the depth at which the active doping profile reaches a concentration of the order of the substrate injection. This has two important implications. First, it means that, in our TP measurements, ΔR_{ac} is only sensitive to the *moderately doped region* of the profile. Second, if the injection level can somehow be changed, the position of the maximum interface peak is shifted accordingly, i.e. part of the profile is scanned. This explains the sensitivity of offset curves to the abruptness of the profiles, as a direct consequence of the lower carrier injection when moving away from the pump beam.

Concerning the DC reflectance R_{dc}, contrary to ΔR_{ac}, Eq. (6.17a) shows that it is mostly sensitive to the *highly doped region* of the profile. It is therefore fairly easy to understand that R_{dc} fixes the peak doping concentration based on the depth and abruptness determined by ΔR_{ac} in the moderately doped region of the profile. This is illustrated in Fig. 7.11, which summarizes the principle of the technique. It is important to mention that, since ΔR_{ac} is blind to the region of the profile situated above a $10^{19}\,\text{cm}^{-3}$ doping concentration, the peak concentration determined by R_{dc} depends on the assumed shape of the profile in that particular region. This causes a uniqueness problem, which already underlines the importance of the quality of the assumed profile shape. This issue is further discussed in Sect. 7.2.3.3

7.2.2 Implementation of the Technique

Based on the theoretical considerations of the previous section, we have developed a technique which, practically speaking, follows this four-step procedure:

- a *substrate* measurement is run, i.e. a measurement on a sample presenting no doped layer but having undergone the same process flow as the sample under investigation. For instance, in the case of an annealed implanted layer, the substrate measurement is run on a region of the sample which has received no dopant implant but has been preamorphized (if relevant) and annealed exactly like the doped sample under investigation. This measurement yields $\Delta R_{\text{ac}}^{\text{Substrate}}$ offset curves and $R_{\text{dc}}^{\text{Substrate}}$,

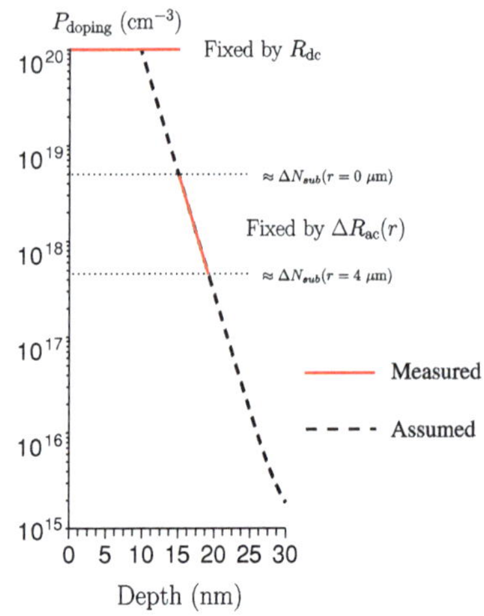

Fig. 7.11 Principle of the profile reconstruction technique. The depth and abruptness at moderate doping are fixed by the measured ΔR_{ac} offset *curves*. The peak doping concentration is fixed by R_{dc} for a certain, assumed, profile shape

- our model for the AC reflectance on homogeneous samples is fitted to the experimental $\Delta R_{ac}^{Substrate}$ offset curves like in Sect. 6.2.1. This determines the excess carrier distribution in the substrate, i.e. $\Delta N_{sub0}(r)$ and $\Delta N_{sub1}(r)$, and the excess temperature $\Delta T_1(r)$,
- a *profile* measurement is run on the sample to be characterized. This yields $\Delta R_{ac}^{Profile}$ offset curves and $R_{dc}^{Profile}$,
- our model for the AC and DC reflectances on non-homogeneous samples is fitted to the experimental $\Delta R_{ac}^{Profile}$ offset curves, using the previously determined $\Delta N_{sub0}(r)$, $\Delta N_{sub1}(r)$ and $\Delta T_1(r)$, and to the experimental $\Delta R_{dc} = R_{dc}^{Profile} - R_{dc}^{Substrate}$.

  It is important to mention that, as explained in Sect. 7.2.1, our measurements offer limited information. Consequently, some constraint on the profile shape has to be imposed (maximum three fitting parameters). In the case of annealed implanted layers, box-like profiles with an exponential tail are a good guess (see e.g. Fig. 1.4), i.e.

$$\begin{aligned} P_{doping}(z) &= N_0 \text{ if } z \leq X_{cst} \\ &= N_0 \times 10^{-(z-X_{cst})/\mathcal{A}} \text{ if } z > X_{cst}, \end{aligned} \tag{7.16}$$

  where N_0 is the peak active doping concentration of the profile, X_{cst} is the depth at which the profile starts to decay and $\mathcal{A}$ is its abruptness (nanometer/decade). The fitting algorithm determines N_0, X_{cst} and $\mathcal{A}$.

Note that the fitting is based on a Levenberg-Marquardt algorithm [2] (minimization of the least-square error).

An important remark concerning the substrate measurement must be made. The technique relies on the substrate measurement for the determination of the substrate plasma and thermal components. As a consequence, the substrate measurement must be carefully monitored. Generalizing Eq. (7.7) to arbitrary profiles, a verification procedure can be used to make sure that the substrate measurement meets the requirements, i.e. $\Delta N_{sub0}(r)$, $\Delta N_{sub1}(r)$ and $\Delta T_1(r)$ are the same in the substrate and profile measurements. In our verification procedure, the substrate measurement $\Delta R_{\text{ac}}^{\text{Substrate}}(r)$ is subtracted from the profile measurement $\Delta R_{\text{ac}}^{\text{Profile}}(r)$ measured on the unknown doped profile. Since the result of this subtraction should eliminate the thermal component, the phase and wavelength of the plasma wave should be recognized. This is illustrated in Fig. 7.12 for samples IIB-D07. The plasma phase and wavelength can be identified in Fig. 7.12d for all layers except the lowest annealing temperature, underlying the thermally dominated behavior of the latter. The technique can safely be used on the other samples since they comply with theory. Interestingly, this also shows that the impact of the boron-implantation-induced damage on the plasma and thermal components is negligible (after anneal).

7.2.3 Discussion of the Results

As can be seen by the multiple fitting algorithms involved in the technique procedure described above, an essential difference with respect to the techniques developed in the previous sections of this chapter is that we here rely fully on our model. We have demonstrated that the dependence of ΔR_{ac} and R_{dc} upon X_j is accounted for with a high accuracy. Unfortunately, we have also seen that the modeled variations in both signals with doping concentration require further improvement. More particularly, it has appeared from our study that the separate fitting of R_{dc} and ΔR_{ac} data on the CVD12 and CVD13 matrices leads to different active doping concentrations (Fig. 6.4). As a consequence of this disagreement, it can be expected that the combined fitting of ΔR_{ac} and R_{dc} data does not converge. However, at sufficiently high doping, the disagreement between both signal modeling should be reduced as a consequence of the vanishing sensitivity of ΔR_{ac} to doping. In other words, given the current status of our model, the technique can only be tested on highly doped layers ($>10^{20}\,\text{cm}^{-3}$). It is important to note that this is *not* a limitation due to the fundamental principle of the technique itself, but to the inaccuracies of our current model.

Taking the last comment into account, the present analysis only looks at B-implanted layers laser annealed with a high temperature. In particular, this section concentrates on the results obtained for samples IIB-D05T5, IIB-D07T5 and IIB-D13T5. These three samples have received the same B implant (energy = 0.5 keV, dose $=10^{15}\,\text{cm}^{-2}$) and anneal (laser annealed three times at 1,300 °C) but different Ge preamorphization implants (PAI), i.e. respectively (5 keV, $10^{14}\,\text{cm}^{-2}$), (5 keV, $5\times10^{14}\,\text{cm}^{-2}$) and (20 keV, $10^{14}\,\text{cm}^{-2}$). The discussion of these three samples is sufficient to reach our final conclusions, which prove to be valid for all the other samples of matrix IIB.

The fitted curves for IIB-D05T5, IIB-D07T5 and IIB-D13T5 are shown in Fig. 7.13a and b for the substrate measurement and (c) and (d) for the profile measure-

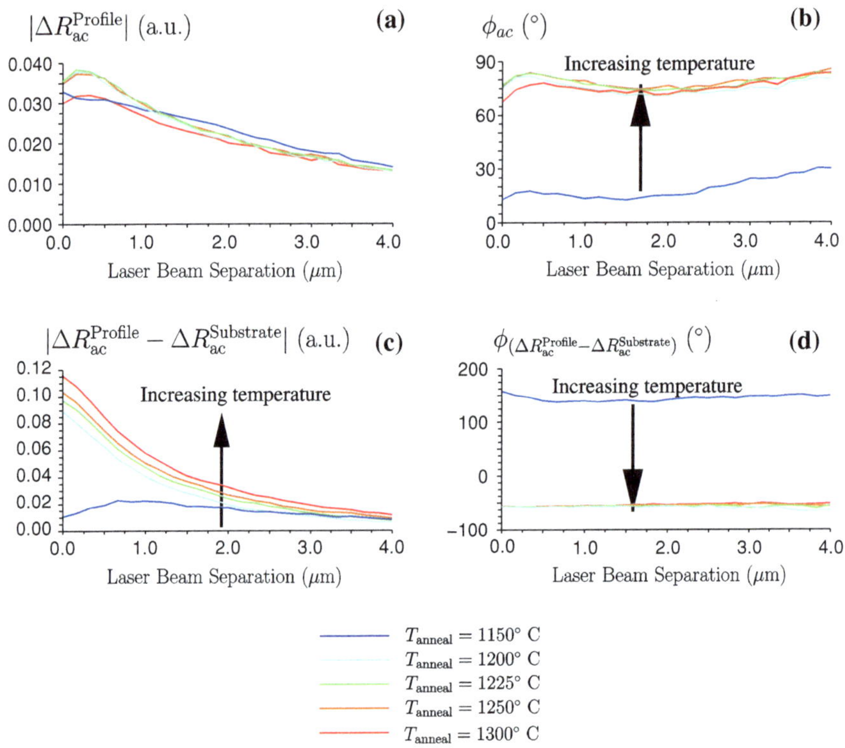

Fig. 7.12 Offset *curves*, i.e. variation with laser beam separation, of **a** the amplitude and **b** the phase of the AC reflectance $\Delta R_{ac}^{Profile}$, **c** the amplitude and **d** the phase of the difference $\left[\Delta R_{ac}^{Profile}(r) - \Delta R_{ac}^{Substrate}(r)\right]$, as measured on samples IIB-D07, i.e. five layers implanted on the same substrate with B (0.5 keV, $10^{15}\,cm^{-2}$) after a Ge preamorphization implant (5 keV, $5 \times 10^{14}\,cm^{-2}$) and laser-annealed three times at 1,150 °C (*blue*), 1,200 °C (*light blue*), 1,225 °C (*green*), 1,250°C (*brown*) and 1,300 °C (*red*). The *arrows* show the trend observed when increasing the annealing temperature. The behavior of the phase and wavelength of the plasma wave are recognized in **d** for the four highest annealing temperatures, which ensures that the technique can safely be used on these samples

ment. It is clearly observed that the theoretical curves nicely fit the experimental data. The so-called *TP profiles* obtained from the fitted data are shown in Fig. 7.13e together with the SIMS profiles measured on the three samples. The SSRM profile measured on sample IIB-D07T5 has also been added (assumed mobility is $100\,cm^2\,V^{-1}\,s^{-1}$). The TP profiles have been obtained in less than 30 min (20-min measurement and <10-min fitting). On average, the quantitative agreement between the TP and active SIMS (and SSRM) profiles is acceptable since the average deviation on the derived depths at a doping of $10^{19}\,cm^{-3}$ is of 3 nm. The average deviation on the peak active doping concentration is of $5 \times 10^{18}\,cm^{-3}$.

However, we note the following important discrepancies. First, the TP profiles are deeper than the SIMS profiles. Second, TP and SIMS are in contradic-

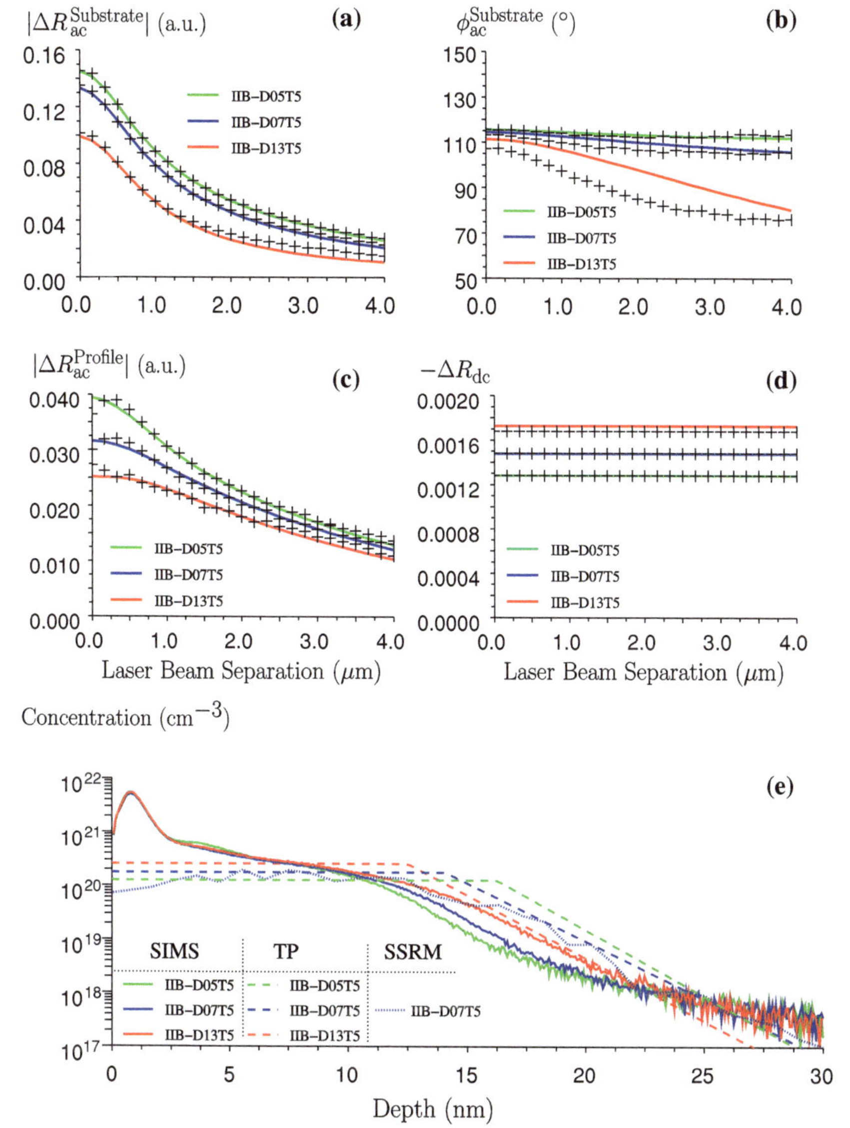

Fig. 7.13 Fitting curves of the experimental **a** $|\Delta R_{ac}^{Substrate}|$, **b** $\phi_{ac}^{Substrate}$, **c** $|\Delta R_{ac}^{Profile}|$ and **d** $\Delta R_{dc} = R_{dc}^{Profile} - R_{dc}^{Substrate}$ measured on samples IIB-D05T5 (*green*), IIB-D07T5 (*blue*) and IIB-D13T5 (*red*), **e** comparison of the TP profiles (*interrupted lines*) with the SIMS profiles (*full lines*) measured on these samples. The SSRM profile of sample IIB-D07T5 is also shown (*dotted line*). The active SIMS doping concentrations of IIB-D05T5, IIB-D07T5 and IIB-D13T5 are respectively 1.71×10^{20}, 1.80×10^{20} and 1.80×10^{20} cm^{-3} (not shown)

tion when it comes to the relative depths of the three profiles. SIMS measures IIB-D05T5 shallower than IIB-D07T5, itself shallower than IIB-D13T5. TP measures exactly the opposite behavior. Note already that this disagreement is quite unexpected since the decrease in $|\Delta R_{ac}^{Profile}|$ from sample IIB-D05T5 down to IIB-D07T13 (Fig. 7.13c) tends to indicate a behavior in agreement with SIMS. We investigate below whether these discrepancies are due to a precision problem inherent to the technique (Sect. 7.2.3.1) or to an accuracy problem due to the modeling error (Sect. 7.2.3.2). Finally, in Sect. 7.2.3.3, the uniqueness of the TP profiles is discussed.

For completeness, as shown in Fig. 7.13, only the amplitude of $\Delta R_{ac}^{Profile}$ is fitted. The phase has indeed been ignored in our fitting algorithm. This is an acceptable approximation for shallow layers where the phase mostly indicates the importance of the thermal component. In this study, the determination of the thermal component is taken care of by fitting the substrate measurement, where the phase is of extreme importance and therefore taken into account (Fig. 7.13b). It can be checked that the experimental phase and the theoretical phase after fitting of the profile measurement are in acceptable agreement.

7.2.3.1 Precision

In order to assess the precision of the developed technique and determine whether the discrepancies observed in Fig. 7.13e are due to random errors, we use the following iterative Monte-Carlo approach. First, at each iteration, we simulate noisy data by adding random noise on top of our experimentally measured data. The noise on the measured $\Delta R_{ac}^{Profile}$ and $R_{dc}^{Profile}$ is assumed to follow a normal distribution, the standard deviation of which is given by the signal repeatability, i.e. respectively $\sigma_{\Delta R_{ac}} = 7 \times 10^{-4}$ (a.u.) and $\sigma_{R_{dc}} = 5 \times 10^{-6}$ (Appendix A.2). To take all possible random errors into account, noise is also added on the substrate measurement data, i.e. $\Delta R_{ac}^{Substrate}$ and $R_{dc}^{Substrate}$. Second, the simulated noisy curves are plugged into our fitting procedure described in Sect. 7.2.2. The fitting values of the three profile parameters N_0, X_{cst} and $\mathcal{A}$ are recorded along the iterations.

Limiting our study to 50 iterations, we obtain the measurement probability distributions shown in Fig. 7.14. The respective precisions σ_{N_0}, $\sigma_{X_{cst}}$ and $\sigma_{\mathcal{A}}$ of the three profile parameters N_0, X_{cst} and $\mathcal{A}$ are defined as the standard deviations of their respective distributions. It can be observed that all three parameters are determined with very high precision. It can also be noted that the peaks of X_{cst} in Fig. 7.14b are clearly separated. As a conclusion, random errors cannot explain the discrepancy between the TP and SIMS profiles observed in Fig. 7.13e.

7.2.3.2 Accuracy

Since the technique proves to determine the profile characteristics with a high precision, it can be expected that the discrepancies between the TP and SIMS profiles are

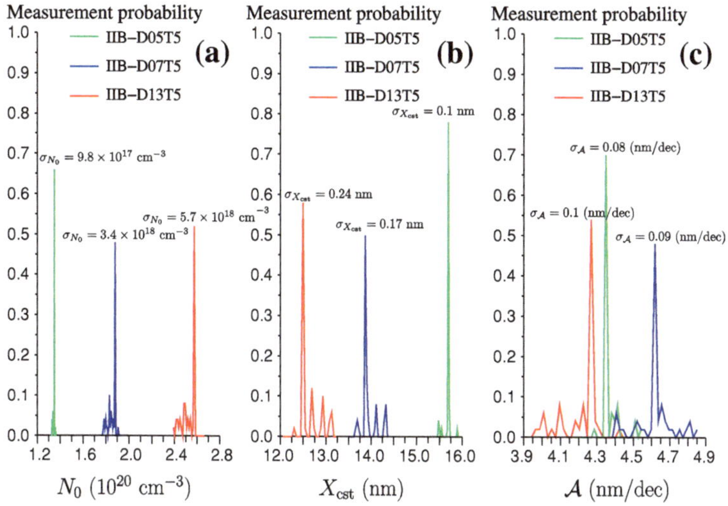

Fig. 7.14 Measurement probability distributions, as obtained from Monte-Carlo simulations, of the three profile parameters i.e. **a** the peak active doping concentration N_0, **b** the depth X_{cst} at which the profile starts to decay and **c** its abruptness $\mathcal{A}$. The narrow peaks show that the developed technique determines all three fitting parameters with very high precision

due to systematic errors and, in particular, to the inadequate accuracy of our model to account for the active doping dependence of the signals.

To prove this statement, we show in Fig. 7.15a the comparison of peak active doping concentration N_0 obtained from our measurement technique with the peak active doping concentration N_{SIMS}^{peak} obtained from SIMS and sheet resistance measurements (bulk mobility assumed). Figure 7.15b shows the comparison of the depths $X_{TP}^{@10^{19}}$ and $X_{SIMS}^{@10^{19}}$ at which respectively the TP and SIMS profiles reach a concentration of $10^{19}\,\text{cm}^{-3}$. Finally, Fig. 7.15c compares the abruptnesses $\mathcal{A}$ and $\mathcal{A}_{SIMS}^{@10^{19}}$ of respectively the TP and SIMS profiles (value taken around a $10^{19}\,\text{cm}^{-3}$ concentration). To underline the sensitivity of the obtained results to the modeling error, we compare the results as obtained with our current model with three modified models, each of which tests the impact of the modeling errors which have been underlined in Sect. 6.3. In particular, Model II shows the impact of a 30 % greater electrorefractive effect, to counteract the observed underestimation of this effect (Sects. 6.3.1 and 6.3.2.1). Model III assumes a doubled modulated irradiance of the pump laser to account for the underestimated excess carrier concentration in the substrate (Sect. 6.3.2.1). Finally, model IV reduces the layer injection, to counteract the overestimated sensitivity to the active doping concentration in the layer (Sect. 6.3.2.2). In particular, the latter model assumes that the constant mode of the excess carrier concentration in the substrate ΔN_{sub0} is equal to $2|\Delta N_{sub1}|$ (lower limit for ΔN_{sub0}). The values used for each of these models are obviously discussable. We believe, however, that they give a good feeling of the impact of the different modeling errors.

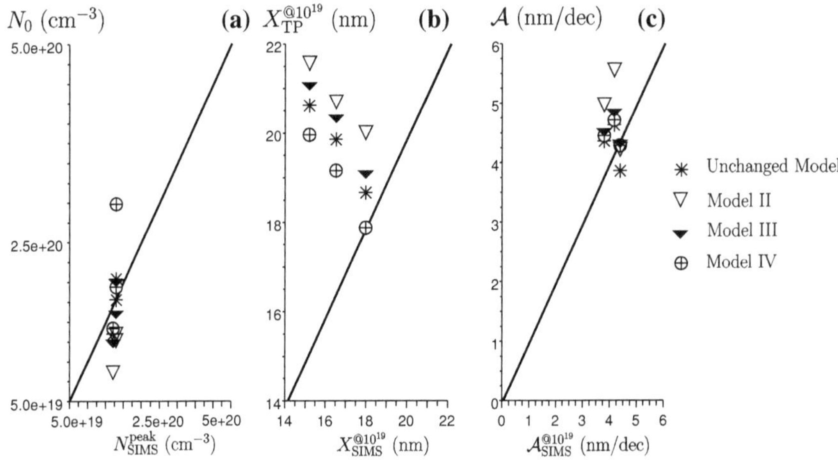

Fig. 7.15 TP versus SIMS profile characteristics and their sensitivity to modeling errors. **a** Comparison of the peak active doping concentration N_0 with the SIMS peak doping concentration N_{SIMS}^{peak} obtained from sheet resistance measurements. **b** Comparison of the depths $X_{TP}^{@10^{19}}$ and $X_{SIMS}^{@10^{19}}$ at which respectively the TP and SIMS profiles reach a 10^{19} cm^{-3} concentration. **c** Comparison of the TP and SIMS profile abruptnesses $\mathcal{A}$ and $\mathcal{A}_{SIMS}^{@10^{19}}$ at a 10^{19} cm^{-3} concentration. For each figure, Model II assumes a 30 % greater electrorefractive effect, Model III assumes a doubled modulated irradiance of the pump laser and Model IV assumes $\Delta N_{sub0} = 2|\Delta N_{sub1}|$

Let us first discuss Fig. 7.15b comparing the depths of the TP and SIMS profiles at a 10^{19} cm^{-3} concentration. As already observed in Fig. 7.13e, the TP profiles are always deeper than the SIMS profiles. Further, the depths of the SIMS and TP profiles are in anti-correlation, i.e. TP measures a deeper profile where SIMS measures it shallower. It can be seen in Fig. 7.15b that, though the depths of the TP profiles vary according to the used model, none of the modified models explains the anti-correlation. The error must therefore be found elsewhere. We believe a likely additional modeling error lies in the substrate measurement fitting. As already mentioned, the evolution of $|\Delta R_{ac}^{Profile}|$ in Fig. 7.13c seems to indicate an evolution of the junction depth in agreement with the SIMS profiles. However, the trend observed in these data is also partly due to the decrease in the substrate plasma component, as can be observed in Fig. 7.13a and as expected when increasing the Ge PAI dose or energy (increased damage). In other words, our model has to discriminate between both effects (reduction in substrate plasma component *and* deeper profile), which it does not seem to achieve accurately. This might indicate that the variations in ΔN_{sub1} and ΔT_1 with the energy and dose of the Ge implant are not properly accounted for. Most likely, the assumed homogeneous ambipolar diffusivity and recombination lifetime have to be blamed.

Moving on to Fig. 7.15c, it can be seen that the abruptness of the TP profiles is in good agreement with SIMS. Furthermore, it shows only limited sensitivity to the used model. The obtained values are therefore quite robust.

Finally, the comparison of the peak active doping concentrations in Fig. 7.15a shows that N_0 is very sensitive to the modeling error. This can easily be explained by the fact that N_0 is fixed by R_{dc} *after* the depth and abruptness have been determined by ΔR_{ac}. Furthermore, the values of N_0 are quite difficult to compare with the N_{SIMS}^{peak} values since the latter assumes crystalline mobility. Finally, the uniqueness of the N_0 value is discussable as studied in Sect. 7.2.3.3 below. To summarize all these comments, the discussion of the impact of modeling error on N_0 is complex and should be considered when the depth accuracy problem is solved.

In conclusion, given the high sensitivity of the depth and active doping concentration of the TP profiles to the modeling error shown in Fig. 7.15, the discrepancies between the TP and SIMS profiles are believed to be due to modeling errors. It seems that, besides the modeling errors highlighted in Sect. 6.3, some of the error would also originate from the fitting of the substrate measurement (impact of Ge implant).

7.2.3.3 Uniqueness

The involvement of a fitting algorithm in the proposed measurement technique raises the question of the uniqueness of the obtained profile. It should, therefore, primarily be checked that the fitting procedure always converges towards the same values of the three profile parameters N_0, X_{cst} and $\mathcal{A}$. Given the discussion of Sect. 7.2.1, however, it appears obvious that the obtained values of the profile parameters are indeed unique. This can, furthermore, be checked by changing the initial guess values of the three parameters in the fitting algorithm. The variations on the final output values are typically of the order of their respective precisions σ_{N_0}, $\sigma_{X_{cst}}$ and $\sigma_{\mathcal{A}}$.

Furthermore, it should also be evaluated how dependent the fitting profiles are on the assumption of their shape. To test this dependence, we propose to compare the TP profiles obtained for samples IIB-D05T5, IIB-D07T5 and IIB-D13T5 when respectively assuming an exponentially decaying profile (Eq. (7.16)) or a profile following the behavior of a complementary error function [3]. Figure 7.16 compares the obtained TP profiles for all three samples. While the obtained depths and abruptnesses in the moderately doped regions are independent from the assumed shape, the peak active doping concentrations are significantly impacted by the profile shape.

This behavior can be explained from the theoretical considerations of Sect. 7.2.1. The depth and abruptness are fixed by the ΔR_{ac} offset curves. Each point of the offset curve being sensitive to a different region of the profile, the moderately doped region of the profile is therefore defined uniquely ($<10^{19}\,cm^{-3}$). The value of R_{dc} is sensitive not only to the peak doping concentration of the profile but to the whole highly doped region ($>10^{19}\,cm^{-3}$). In other words, the determination of the highly doped regions relies on one single value of R_{dc} which leads to the uncertainty shown in Fig. 7.16.

Note that, in order to eliminate the ambiguity, the measurement of ΔR_{ac} with higher substrate injection should be considered.

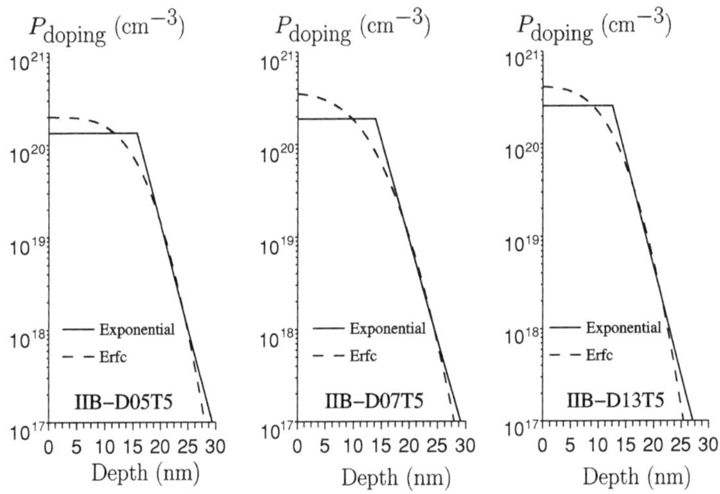

Fig. 7.16 Comparison of the TP profiles as obtained from the fitting of the experimental data measured on **a** IIB-D05T5, **b** IIB-D07T5 and **c** IIB-D13T5 assuming a profile decaying exponentially (*full lines*) or a profile following a complementary error function (*interrupted lines*). While the depth and abruptness are unique, the peak active doping concentration depends upon the profile shape

In summary, we have developed a fast non-destructive profile characterization technique based on the combined measurement of ΔR_{ac} offset curves and R_{dc}. The technique has a high precision and is in acceptable agreement with SIMS and SSRM. It, however, suffers significantly from the modeling errors highlighted in Sect. 6.3 as well as extra modeling error induced by the fitting of the substrate measurement (impact of Ge implant). Further, the depth and abruptness of the profiles obtained using this technique are unique but the obtained peak active doping concentration depends on the assumed profile shape. A higher pump irradiance should solve this uniqueness problem.

7.3 Summary

In this chapter, we have developed three fast non-destructive techniques to derive information about active doping profiles from TP measurements. Two of these techniques are based on the use of ΔR_{ac} and focus on the determination of junction depths of box-like active doping profiles. The third technique combines the use of ΔR_{ac} and R_{dc} for the complete characterization of active doping profiles.

The first junction depth determination method, developed in Sect. 7.1.1, is based on the measurement of ΔR_{ac} offset curves. Equation (7.3) demonstrates that, owing to the fast lateral decay in the layer plasma and thermal components, the junction depth of box-like profiles should indeed be directly accessible in ΔR_{ac} when the pump and probe beams are sufficiently spaced apart. Practically speaking, however, experimental data have shown that the layer plasma component does not decay fast enough (Fig. 7.4). As a result, it is very unlikely that high precision and accuracy can be reached simultaneously with this technique.

The second junction depth determination technique, developed in Sect. 7.1.2, is based on the efficient combination of the ΔR_{ac} measured respectively on a homogeneously doped substrate and on a reference doped layer. Equation (7.8) shows that these two extra measurements are sufficient to cancel the layer plasma and thermal components and therefore lead to the junction depth. The technique has been successfully applied and, thanks to a sub-nanometer precision and accuracy almost independently from junction depth, its agreement with SIMS is excellent (Fig. 7.6). We believe this technique has a very high applicability potential for the test of depth uniformity over full wafers. Layers in the range of $X_j \approx 44$ nm should however be avoided (Figs. 7.7, 7.8).

Finally, the complete profile characterization technique, developed in Sect. 7.2, is based on the combination of the ΔR_{ac} offset curves and R_{dc}. We have proven that this combination allows for the simultaneous measurement of the peak active doping concentration, depth and abruptness of active doping profiles. The technique has been proven to be highly precise but to suffer significantly from the insufficient accuracy of our model highlighted in Sect. 6.3 as well as the inaccurate modeling of the impact of Ge-implantation-induced damage. The uniqueness of the obtained profiles requires either an a priori knowledge of the profile shape or a higher pump irradiance.

References

1. J. Bogdanowicz, F. Dortu, T. Clarysse, W. Vandervorst, E. Rosseel, N.D. Nguyen, D. Shaughnessy, A. Salnik, L. Nicolaides, Nondestructive extraction of junction depths of active doping profiles from photomodulated optical reflectance offset curves. J. Vac. Sci. Technol. B **28**(1), C1C1–C1C7 (2010)
2. M. Heath, *Scientific Computing, an Introductory Survey* (McGraw-Hill, New York, 1997)
3. M. Abramowitz, I. Stegun, *Handbook of Mathematical Functions* (Dover, New York, 1964)

Chapter 8
Conclusions and Recommendations

In this work, we have (i) developed, (ii) tested and (iii) made use of a physical model accounting for the behavior of high-frequency Photomodulated Optical Reflectance (PMOR) on homogeneously and non-homogeneously doped silicon. Sections 8.1–8.3 below summarize the most important results of each of these tasks. In Sect. 8.4, we give an outlook on future work.

8.1 Development of the Model

The model has been constructed from Chaps. 2–5.

To understand PMOR, the first question to be answered concerns the physical origin of a change in reflectance. Chapter 2 has shown that a modification in the reflectance of a sample is the consequence of the perturbation of its complex refractive index. Very useful analytical relationships between the perturbation of the reflectance and the perturbation of the complex refractive index have been derived. Importantly, these expressions show that the perturbation of the reflectance depends on the depth-profile of the complex refractive index perturbation.

Chapter 3 has explained the physical phenomena underlying a perturbation of the complex refractive index. Though numerous phenomena can induce such a perturbation, Chap. 3 has proven that only free-carrier and thermal effects have to be taken into account. In other words, it is a change in free carrier concentration and/or temperature which generates the refractive index perturbation and, hence, the reflectance perturbation which we observe and measure. It has also been shown that only real perturbations of the refractive index are to be expected in the framework of our investigations. A very simple final expression linking the perturbation of the real refractive index and the excess carrier concentration and temperature has been derived.

Chapter 4 has studied quantitatively how the pump and probe lasers induce the presence of excess carriers and temperature in a homogeneously doped silicon sample. This has required deriving and solving carrier and heat transport equations. This chapter has initially started with the complete thermodynamic model, i.e. a coupled

J. Bogdanowicz, *Photomodulated Optical Reflectance*, Springer Theses,
DOI: 10.1007/978-3-642-30108-7_8, © Springer-Verlag Berlin Heidelberg 2012

system of four nonlinear time-dependent equations involving all the possible transport and generation mechanisms of both free carriers and heat. These four equations solve for the time-dependent electric-field, electron, hole and temperature distributions. An in-depth two-step analysis has allowed to strongly simplify these equations in the specific context of our investigations, i.e. TP in silicon. First, we have shown that the transport problem can be simplified into two equations solving respectively for the excess carrier distribution, i.e. the equal excess electron and hole distributions, and for the excess temperature distribution. Second, switching to the Fourier domain, we have derived the steady-periodic form of our transport equations, i.e. the equations solving separately for each Fourier coefficient of the excess carrier and temperature distributions. The final coupled system of four steady-periodic nonlinear equations solve for the constant and fundamental modes of the excess carrier and temperature distributions. These final transport equations have been solved, demonstrating that the constant and fundamental modes of each distribution behave respectively as decaying exponentials and damped waves. The characteristics of these decaying exponentials, i.e. their amplitudes and decay lengths, and of these damped waves, i.e their amplitudes, decay lengths, phases and wavelengths, have been calculated as a function of the doping concentration of the silicon sample.

To conclude our model, Chap. 5 has extended the carrier and heat transport theory to non-homogeneously doped silicon and, more particularly, to the case of ultra-shallow highly doped layers on top of lowly doped silicon substrates. Two approximations, as demonstrated, suffice to strongly simplify this problem. First, we have assumed that the quasi-Fermi levels of the electrons and holes are flat through the space-charge region. Based on this assumption, we have derived an algebraic expression linking the excess carrier profile in a doped layer to the active doping profile in this layer and the excess carrier concentration present in the substrate region, i.e. below the doped layer. We have shown that this approximation is valid up to very high active doping concentrations in the layer and, as a consequence, that it can be used independently from this concentration. Second, the excess carrier and temperature distributions in the substrate have been assumed not to be impacted in any way by the presence of the doped layer, i.e. they have been assumed to follow exactly the theory developed in Chap. 4. We have shown that this is a very good approximation for the temperature distribution and for the fundamental mode of the excess carrier distribution. The constant mode of the excess carrier distribution, however, has proven to be influenced by the presence of the doped layer, more particularly in the case of highly doped p-type layers.

8.2 Evaluation of the Model

The construction of a model always requires testing against experimental data. For this purpose, Chap. 6 has evaluated the ability of the developed model to explain the experimental data measured respectively on homogeneously doped and non-homogeneously doped silicon with the TP630XP tool.

Final model equations summing up our model for R_{dc} and ΔR_{ac} have been derived respectively for homogeneously doped silicon and box-like active doping profiles. The model equations for R_{dc} show its linear variation with the active doping concentration, whether or not it be uniform, and its cosinusoidal dependence upon junction depth (when relevant). The model equations for ΔR_{ac} are more complex. On homogeneously doped silicon, ΔR_{ac} consists of two components, i.e. the plasma and thermal components, respectively linked to the fundamental modes of the excess carrier and temperature distributions at the sample surface. Both components vary with the doping concentration of the substrate but have opposite signs. On box-like active doping profiles, ΔR_{ac} is composed of three components, i.e. the substrate plasma, layer plasma and thermal components, respectively linked to the fundamental mode of the excess carrier distribution in the substrate, the fundamental mode of the excess carrier distribution in the layer and the fundamental mode of the temperature distribution. While the substrate plasma component only varies with the junction depth of the profile, the layer plasma component changes with both the junction depth and the active doping concentration of the doped layer. More particularly, the dependence of the latter component upon doping concentration drops at high doping concentration. The thermal component is independent from the doped layer.

The quality of our model equations has been evaluated against TP measurements run on homogeneously doped silicon and Chemical-Vapor-Deposited (CVD) box-like profiles. Though our model has shown its ability to account with very high accuracy for the variations in signals with junction depth, some discrepancies have been observed on their dependence upon active doping concentration in doped layers. These discrepancies are recapitulated in Sect. 8.4 below.

8.3 Application of the Model

Based on the developed and tested model, Chap. 7 has finally evaluated the capabilities of PMOR for fast and non-destructive carrier-depth profiling in silicon. Two techniques have been developed for the determination of the junction depth of box-like active doping profiles based on the measurement of ΔR_{ac}. The first technique relies on the measurement of the ΔR_{ac} offset curves, wherein ΔR_{ac} is measured as a function of the probe and pump beam separation, respectively on the unknown box-like doping profile and on a homogeneously doped substrate as found below the doped layer. We have shown that separating the pump and probe beams should, in theory, eliminate the layer plasma and thermal components and therefore leave the junction depth as sole dependence. In practice, however, due to a residual layer plasma component, this technique is very unlikely to be applicable in practice.

The second technique relies on the combination of the AC reflectances measured on the unknown box-like profile, on a homogeneously doped substrate as found below the doped layer and on a second doped layer with a known junction depth and approximately the same doping concentration as the unknown layer. This second technique has proven to be in excellent correlation with SIMS thanks to simultaneous sub-nanometer precision and accuracy. The high accuracy of this technique is based on the fact that the dependence of ΔR_{ac} upon doping has been canceled. This technique has a high applicability potential for the fast and non-destructive determination of the relative variation in junction depth over full wafers. Note that, very interestingly, this application can be performed with the TP630XP tool in its current implementation.

Last and most important of all, a model-based technique relying on the simultaneous fitting of ΔR_{ac} offset curves and R_{dc} has been developed for the complete characterization of active doping profiles. We have shown that ΔR_{ac} offset curves are not only sensitive to the depth of profiles but also their abruptness. Thanks to the high sensitivity of R_{dc} to the peak active doping concentration, as demonstrated, the combined use of ΔR_{ac} offset curves and R_{dc} allows for the complete reconstruction of profiles. The technique has been tested on low-energy high-dose annealed B-implanted layers with different Ge PAI conditions. The observed average deviation versus SIMS on the depths of the profiles at a concentration of $10^{19}\,\mathrm{cm}^{-3}$ is of 3 nm. The average deviation on the peak active doping concentration is of $5 \times 10^{18}\,\mathrm{cm}^{-3}$ (bulk mobility assumed for the active SIMS concentration). This final development takes PMOR very close to the ideal technique sought in the Introduction of this work. More particularly, this technique fulfills all the requirements mentioned in Sect. 1.1, except for its accuracy, which needs further improvement due to the remaining modeling errors. As a consequence, we propose to recapitulate our recommendations for an improved accuracy of the technique in Sect. 8.4 below.

8.4 Future Work

We believe future fundamental work is needed on three aspects of our model. First, the electrorefractive effect should be better quantified. Second, the quantitative understanding of the behavior of ΔR_{ac} on homogeneously doped substrate should be improved. Finally, the quantification of the excess carrier concentration in doped layers should be enhanced.

Starting with the electrorefractive effect, both ΔR_{ac} and R_{dc} data have shown that this effect is underestimated by our model. To solve this issue, Hall measurements should be run on the CVD12 and CVD13 matrices so as to determine the actual active doping concentrations in these layers. Combining these values to the R_{dc} measured on these layers would offer a direct measurement of the accurate value of $\partial n/\partial P$. The generation and measurement of an equivalent matrix of n-type layers would lead to the determination of an accurate value of $\partial n/\partial N$.

The work needed for the better understanding of the behavior of ΔR_{ac} on homogeneously doped substrate is not as straightforward and consists of two different tasks. First, our calculations appear to underestimate the excess carrier concentration in lowly doped substrates most likely due to an underestimated band-to-band absorption coefficient and overestimated ambipolar diffusivity. Extra work is therefore needed to model the variations with injection in these two parameters. Note that bandgap narrowing is involved in the modeling of both parameters, which highlights the importance of the latter effect. Second, as noted in Sect. 7.2.3, more fundamental work is also needed on annealed preamorphized substrates. The impacts of the preamorphization and of the annealing should be investigated separately and better understood. Some preliminary results are given in Appendix A.4.

Finally, an improved modeling of the excess carrier concentration in doped layers would, first, require more accurate quantification of the bandgap narrowing effect. Second, the inclusion of the impact of the doped layer on the excess carrier distribution in the substrate would also be required. Let us recall that a simple solution to this last issue lies in an increase in pump irradiance (Chap. 5).

As a final word, let us discuss some of the tool modifications which could lead to an enhanced technique. First and foremost, the technique would benefit in multiple ways from an increased pump irradiance. First, as noted above, a higher pump irradiance would render the substrate excess carrier distribution fully independent from the doped layer and therefore reduce significantly the modeling errors. Based on the comparison of Fig. 5.3a, c, it appears that an increase in pump irradiance by a factor $\simeq 2$ should be sufficient. Second, a higher pump irradiance would lead ΔR_{ac} to being sensitive to the more highly doped regions of the profile. As a consequence, it would solve the uniqueness problem highlighted in Sect. 7.2.3.3. Based on Fig. 7.16, it appears that the regions up to $2 - 3 \times 10^{19}\,\text{cm}^{-3}$ are still uniquely defined by the technique. In other words, following a linear reasoning, an increase in pump irradiance by a factor $\simeq 5$ should allow for unique determination of the profile in regions doped up to $\simeq 10^{20}\,\text{cm}^{-3}$. This would strongly reduce the uncertainty on the peak doping concentration fixed by the DC reflectance.

Second, we believe that there is little chance that a larger beam separation would help derive more information about the profiles. Based on Fig. 7.16, it seems that the doping range which the current ΔR_{ac} offset curves are sensitive to goes down to quite low doping concentrations ($<10^{18}\,\text{cm}^{-3}$). The more lowly doped regions are not of very high interest. Besides, larger beam separations would lead to very low signal levels, i.e. low precision.

Third, a change in the modulation frequency of the pump irradiance might lead to some improvement of the technique. However, advantages as well as drawbacks of lower and higher modulation frequencies can be found. Going to lower modulation frequencies would advantageously lead to a higher carrier injection but would simultaneously increase the sensitivity to recombinations. A higher modulation frequency would shorten the thermal decay length and therefore enhance the sensitivity of the signals to the plasma components at large beam separation. However, it would also reduce the excess carrier injection. In other words, it is difficult to evaluate which direction should be taken.

Fourth, similar conclusions can be reached concerning the pump and probe laser wavelengths. We here only mention the drawbacks. Concerning the pump laser wavelength, on the one hand, its increase would disadvantageously lead to a reduced absorption coefficient, i.e. lower carrier injection. On the other hand, a shorter pump wavelength would increase the direct heating and therefore the thermal component.

As for the probe laser, reducing its wavelength would lead to a reduced Drude electrorefractive effect and therefore lower plasma components. A longer probe wavelength would lead to reduced sensitivity to junction depth.

In conclusion, we believe that an increase in the pump irradiance as well as the consideration of the mentioned fundamental issues would lead PMOR to fulfilling all the requirements for the non-destructive carrier-profiling tool sought in our Introduction.

Appendix

A.1 Measurement Procedure

This Appendix introduces the procedure which we have developed for the measurements presented in this work. In particular, in the framework of our investigations, we have introduced two improvements to be discussed. First, Sect. A.1.1 demonstrates that a measurement should be run for a long enough period of time so as to have a high repeatability. Second, Sect. A.1.2 shows that, in order to ensure the commensurability of the signals from one wafer to another, the quality of the substrate should be very carefully monitored. As a consequence of these two observations, every data point shown in this work corresponds to the signal values after the pump and probe have shone on the measurement spot for 25 s. Further, all the used CVD layers have been grown on device-quality wafers.

A.1.1 Surface Charging

As already pointed out and discussed in Refs. [1, 2], the AC reflectance ΔR_{ac} varies with illumination time, leading to very poor signal repeatability. The deeply documented and argumented conclusions of Ref. [2] demonstrate that this effect is due to a variation in the SRV with illumination time, as a result of charge injection into the surface oxide. In particular, at the moderate pump irradiance of the TP pump laser, the AC reflectance is expected to *increase* with time, as a result of the *reduction* in SRV and, therefore, of an *increase* in the plasma component.

These theoretical investigations are in agreement with the experimental charging curves observed in Fig. A.1a and b, i.e. the variations in the amplitude and phase of the AC reflectance observed when measuring these signals as a function of time on the same spot of a sample. These Figures have been obtained by measuring ΔR_{ac} seven consecutive times for 100 s with aligned beams (zero separation) on a lowly-doped n-type device-quality substrate. The pump and probe beams are shut for a few seconds in between each round of 100 s (limitation of our measurement

J. Bogdanowicz, *Photomodulated Optical Reflectance*, Springer Theses,
DOI: 10.1007/978-3-642-30108-7, © Springer-Verlag Berlin Heidelberg 2012

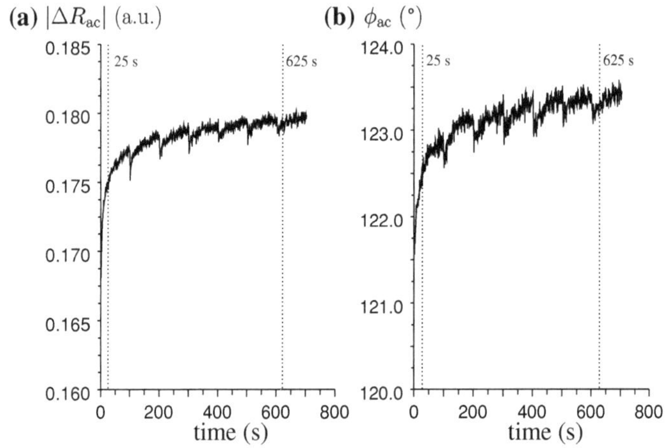

Fig. A.1 Charging *curves*: variation with time in **a** the amplitude and **b** the phase of the AC reflectance measured on a device-quality lowly doped n-type wafer. A steep increase in amplitude and phase is initially observed before both start to saturate. The sharp drop in amplitude and phase observed every 100 s are due to the pump and probe lasers being shut every 100 s. The first data point of each offset curve measured in this work has been recorded after 25 s and the 25th and last one after 625 s as highlighted by the *vertical dotted lines*

routine), which explains the sharp drop in amplitude and phase every 100 s. Two observations are of the utmost importance in Fig. A.1. First, the variation in signals in the first seconds of the measurement is very steep, leading to a very poor signal repeatability if one measures in that region. Second, the variations tend to saturate after a certain time. This corresponds to the saturation of the SRV discussed in Ref. [2]. However, even after 700 s, the signal is not fully saturated.

The measurement procedure we have developed is based on these two observations. The first observation pushes us to shine the pump and probe lasers for a certain while on the sample before actually recording the AC reflectance, in order to maintain a high repeatability. The second observation determines after how long we should record ΔR_{ac}. Ideally, one should measure at saturation, i.e. after more than 700 s. However, the measurement time then becomes prohibitive. One therefore has to compromise between acceptable measurement time and repeatability. For the measurements shown in this work, all data points have been obtained after shining the pump and probe beams for **25 s** on the measurement spot. This might not seem sufficient, considering the charging curves of Fig. A.1. However, we always measure offset curves in this work. In particular, each offset curve we have measured contain 25 data points measured with the probe constantly shining on the same spot and the pump laser shining on a nearby spot (maximum 4 μm away). The last data point is therefore recorded after approximately 625 s, i.e. almost at saturation. Though the first data points of our offset curves obviously still bear to some extent the trace of the charging effect, this measurement procedure allows us to measure repeatable offset curves in a reasonably short period (≈10 min).

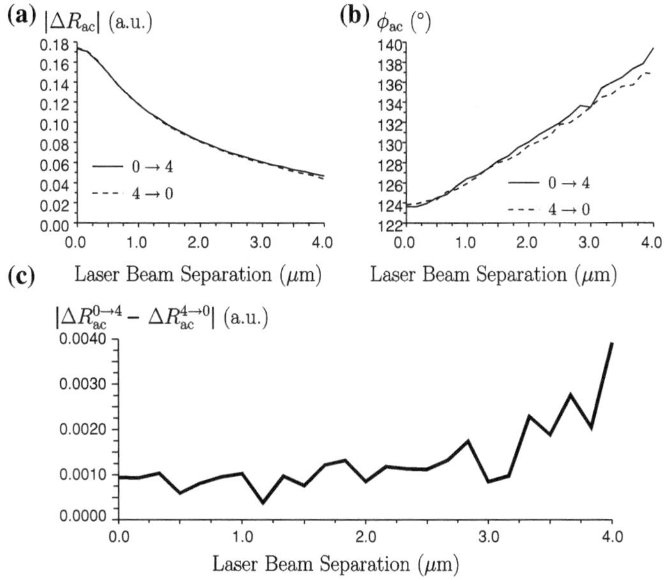

Fig. A.2 Comparison of offset *curves* of **a** the amplitude and **b** the phase of ΔR_{ac} as measured on a device-quality lowly doped n-type substrate when the beams are initially aligned (0 → 4) or initially at their maximum separation (4 → 0). **c** Amplitude of the difference between each offset *curve*. The difference is of the order of the signal repeatability

To confirm this statement, we compare in Fig. A.2a and b two amplitude and phase offset curves obtained on the same substrate following our procedure. One of the curves has been measured starting with the beams aligned initially and *increasing their separation* of 0.167 μm every 25 s (0 → 4). The second offset curve has been measured on the same spot starting with an initial beam separation of 4 μm and *decreasing their separation* of 0.167 μm every 25 s (4 → 0). The curves are very similar. This is corroborated by Fig. A.2c where the amplitude of the difference of both offset curves proves to be of the order of the signal repeatability (See Sect. A.2 below). As a final remark, note that, as highlighted by Fig. A.2c, the signals at large beam separations are more impacted by the charging effect. As a consequence, all our offset curves are measured from 0 to 4 μm.

In summary, our measurement procedure is as follows:

- each measured data point is acquired after shining the pump and probe beams for 25 s on the measured spot,
- each offset curve contains 25 points, spaced apart of a distance of 0.167 μm,
- each offset curve is measured with, initially, aligned beams and increasing beam separation (0 → 4).

  Note that, as an additional benefit, the measurement procedure also reduces the sensitivity of our measurements to SRV. The impact of SRV has actually not been observed in our measurements.

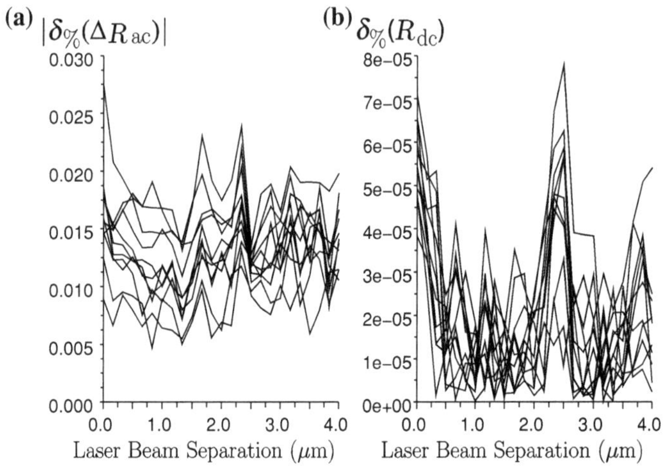

Fig. A.3 Relative wafer-to-wafer variation observed on the offset curves of **a** ΔR_{ac} and **b** R_{dc} measured before growth of the CVD layers on the device-quality substrates of the CVD12 matrix. The relative variation in ΔR_{ac} and R_{dc} being respectively lower than 2 and 0.01 %, the variations in signal observed after growth of the layers (Fig. 6.3) are due to the doped layers themselves and not to the underlying substrates

A.1.2 Quality of the Substrates

The consideration of surface charging allows us to measure with a high repeatability on a specific spot of a specific wafer. However, this work also requires that the measurements should be reproducible from one wafer to another. Ideally, the substrates below each CVD layer we have used to develop our model should be the same, so as to make sure that the substrate plasma and thermal components are equal for all the CVD layers of our database. This ensures that the signal variations we observe on the different CVD layers are due to a thickness or a layer active doping difference, and not to a substrate difference.

As a consequence, monitor-quality wafers should be avoided. On n-type monitor wafers, we have noted variations in ΔR_{ac} larger than 10 % from one substrate to the other (see e.g. Fig. 5 of Ref. [3]). This is of course unacceptable when one wants to build a quantitative theory. On the contrary, Fig. A.3a shows that the relative wafer-to-wafer variations in the ΔR_{ac} offset curves measured on the device-quality n-type substrates of the CVD12 matrix before growth of the CVD layers are below $\approx 2\,\%$. Similarly, the relative wafer-to-wafer variations in R_{dc} are below 0.01 %. In other words, the signal variations observed on the CVD12 matrix after growth of the layers are confirmed to be due to the doped layers themselves.

This is the second improvement of our measurement procedure. We recommend that only device-quality wafers should be used. Besides, we recommend a measurement should be run on the substrate before growth or

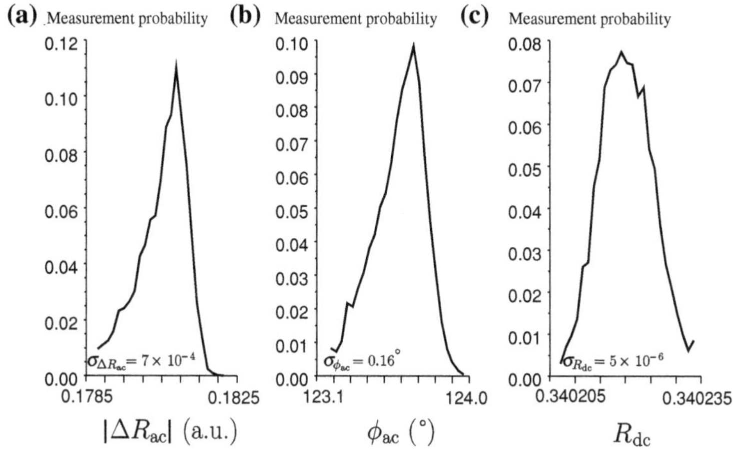

Fig. A.4 Measurement repeatability: measurement probability distributions of **a** $|\Delta R_{ac}|$, **b** ϕ_{ac} and R_{dc} as obtained when measuring 5,000 times on the same spot a device-quality lowly doped n-type substrate (aligned beams). A 300 s charging period has been used

implantation of the doped layer so as to ensure that the substrate behaves as expected. This greatly simplifies the understanding of the signals after growth/implantation.

Note that the physical origin of the large wafer-to-wafer variations in signals observed on monitor-quality substrates is still under investigation.

A.2 Measurement Repeatability

The main benefit of our measurement procedure (Appendix A.1) is to enhance the signal repeatability. In order to evaluate the actual values of the signal repeatabilities, we have measured ΔR_{ac} and R_{dc} 5,000 times on the same spot of a device-quality lowly doped n-type substrate (with aligned beams). In order to account for the impact of charging in our measurement procedure, the surface has been charged 300 s before starting to record the signal values.

The measurement probability distributions of respectively the amplitude of the AC reflectance, the phase of the AC reflectance and the DC reflectance are shown in Fig. A.4a–c. All three distributions are sharply peaked. Notice that the distributions of the amplitude and phase of the AC reflectance are not perfectly symmetric as a result of charging, still present after 300 s. The standard deviation of each distribution gives the measurement repeatability. Most importantly, we have $\sigma_{\Delta Rac} = 7 \times 10^{-4}$ (a.u.) and $\sigma_{Rdc} = 5 \times 10^{-6}$. These excellent repeatability values are used in this work for the determination of the precision of the developed techniques (Chap. 7).

Note that, though the obtained $\sigma_{\Delta Rac}$ contain some contribution of the charging effect, it is quite robust. For example, $\sigma_{\Delta Rac} = 7.9 \times 10^{-4}$ (a.u.) is obtained when, simultaneously, the initial charging period is limited to 100 s and the total number of measurements (and hence total time) is reduced to 1,000. As a consequence, $\sigma_{\Delta Rac} = 7 \times 10^{-4}$ (a.u.) is believed to represent accurately the repeatability of $|\Delta R_{\mathrm{ac}}|$ when using our measurement procedure.

A.3 Experimental Database

A.3.1 Homogeneously Doped Silicon

Our matrix of p- and n-type homogeneously doped substrates is described in Table A.1.

A.3.2 Box-Like Chemical-Vapor-Deposition Layers: CVD12 and CVD13 Matrices

All the CVD layers used in this work were grown with an ASM epsilon 2,000 tool on 200 mm device-quality substrates. Both the CVD12 (Table A.2) and CVD13 (Table A.3) matrices are B-doped.

A.3.2.1 CVD12 and CVD13 Matrices: SIMS Profiles

The SIMS profiles of the complete CVD12 and CVD13 matrices are shown in Figs. A.5 and A.6. Only the ^{11}B concentrations are shown, which corresponds to 80.1 % of the total B concentration (19.9 % ^{10}B to be added). The SIMS profiles have been measured in the center of each wafer. While the SIMS profiles of all the layers of the CVD12 matrix have been measured, only half of the CVD13 layers have been measured (layers with labels x.2, x.4, x.6 with x =1, 2, 3). The junction depths of x.1, x.3 and x.5 (x = 1, 2, 3) have been linearly interpolated based on the junction depths of x.2, x.4 and x.6 (x = 1, 2, 3). This linear increase in junction depth from x.1 to x.6 is expected since the deposition time has also been increased linearly during their growth.

A.3.2.2 CVD12 and CVD13 Matrices: Active Doping Concentrations

Figure A.7 illustrates the method we have used to determine the active doping concentrations of our CVD layers, following Ref. [4]. The inverse of the sheet resistance (measured here with RsL) is plotted vs the SIMS junction depth

Table A.1 Description of our matrix of homogeneously doped substrates

Label	Doping type	Resisitivity (ohm.cm)	Act. doping conc. (cm^{-3})
SUB1	p-type	1.56×10^{-2}	4.62×10^{18}
SUB2	p-type	1.14×10^{-2}	7.10×10^{18}
SUB3	p-type	1.28×10^{-2}	5.44×10^{18}
SUB4	p-type	1.08×10^{-2}	7.71×10^{18}
SUB5	p-type	8.04×10^{-3}	1.13×10^{19}
SUB7	p-type	2.49	5.54×10^{15}
SUB8	p-type	2.17	6.39×10^{15}
SUB9	p-type	9.08×10^{-1}	1.62×10^{16}
SUB103	p-type	1.03	1.41×10^{16}
SUBP10	p-type	10	1.34×10^{15}
SUBP5060	p-type	50-60	$\approx 2.2\times10^{14}$
SUBP01	p-type	1.0×10^{-1}	2.77×10^{17}
SUBP001	p-type	1.0×10^{-2}	8.49×10^{18}
SUBP478	p-type	4.78	2.83×10^{15}
SUBP234	p-type	23.4	5.7×10^{14}
SUBP004	p-type	4.2×10^{-2}	8.41×10^{17}
SUBP510m	p-type	9.33×10^{-3}	8.85×10^{18}
SUBP15m	p-type	3.89×10^{-3}	2.79×10^{19}
SUB6	n-type	1.76×10^{-3}	3.92×10^{19}
SUBN092	n-type	0.9-2	$\approx 4\times10^{15}$

Table A.2 Description of the CVD12 matrix

Label	$X_j^{@90\%}$ (nm)[a]	R_s (ohm/sq)[b]	Layer Act. doping conc. (cm^{-3})[c]
CVD 12 matirx			
1.1	7.2	3700	6.4×10^{19}
1.2	16.8	1176	6.4×10^{19}
1.3	28.0	696	6.4×10^{19}
1.4	38.2	490	6.4×10^{19}
1.5	47.7	380	6.4×10^{19}
1.6	59.8	311	6.4×10^{19}
2.1	4.6	N/A	1.3×10^{19}
2.2	15.2	8850	1.3×10^{19}
2.3	25.1	3960	1.3×10^{19}
2.4	35.1	2478	1.3×10^{19}
2.5	46.9	1810	1.3×10^{19}
2.6	56.8	1420	1.3×10^{19}

[a] The SIMS junction depth of CVD box-like profiles is defined as the depth at which the SIMS profile reaches 90% of its maximum concentration

[b] The values of the sheet resistance have been measured with RsL

[c] The values of the active doping concentration have been obtained from the slope of the $1/R_s(X_j^{@90\%})$ curve (Appendix A.3.2.2)

Table A.3 Description of the CVD13 matrix

Label	$X_j^{@90\%}$ (nm) [a]	R_s (ohm/sq)[b]	Layer Act. doping conc.(cm^{-3})[c]
CVD 13 matirx			
1.1	3.9[d]	8477	4.5×10^{19}
1.2	15.6	1845	4.5×10^{19}
1.3	25.7[e]	1020	4.5×10^{19}
1.4	34.9	693	4.5×10^{19}
1.5	47.5[e]	526	4.5×10^{19}
1.6	59.2	434	4.5×10^{19}
2.1	5.7[d]	27436	3.3×10^{19}
2.2	15.9	3328	3.3×10^{19}
2.3	25.4[e]	1678	3.3×10^{19}
2.4	34.5	1114	3.3×10^{19}
2.5	45.0[e]	816	3.3×10^{19}
2.6	55.2	661	3.3×10^{19}
3.1	4.4[d]	32202	1.2×10^{19}
3.2	14.4	5504	1.2×10^{19}
3.3	24.3[e]	3157	1.2×10^{19}
3.4	34.0	2250	1.2×10^{19}
3.5	44.2[e]	1740	1.2×10^{19}
3.6	54.2	1405	1.2×10^{19}

[a] The SIMS junction depth of CVD box-like profiles is defined as the depth at which the SIMS profile reaches 90% of its maximum concentration

[b] The values of the sheet resistance have been measured with RsL

[c] The values of the active doping concentration have been obtained from the slope of the $1/R_s(X_j^{@90\%})$ curve (Appendix A.3.2.2)

[d] Value linearly extrapolated from the three actually measured $X_j^{@90\%}$ of the set

[e] Value linearly interpolated from the three actually measured $X_j^{@90\%}$ of the set

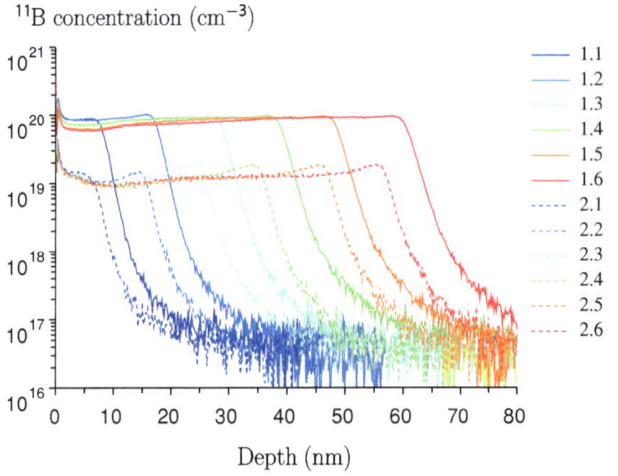

Fig. A.5 SIMS profiles of the ^{11}B content (abundance = 80.1 %) in the layers of the CVD12 matrix

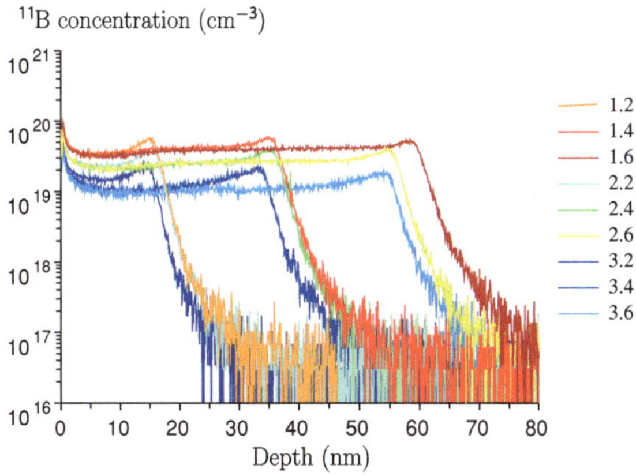

Fig. A.6 SIMS profiles of the ^{11}B content (abundance = 80.1 %) in 9 of the 18 layers of the CVD13 matrix

(at 90 % of the peak concentration) for each of the five sets of different active doping concentrations. The slopes of the obtained curves give the conductivity of the layer in ohm $^{-1}$nm^{-1}. Using Klaassen's mbility model [5], i.e. assuming that the mobility of the majority carriers in the layer is the same as in a bulk silicon sample of the same active doping concentration, a value of the active doping concentration can be derived. This value is actually a lower limit as mentioned in Sect. 1.2.1.3.

Note that each interpolation curve crosses the x-axis (i.e. $1/R_s = 0$) at a positive value. Whether this feature is due to the presence of an inactive layer [4] or a depletion effect [6] is still under discussion. Note that the crossing value typically increases for decreasing active doping concentration [4]. This is not strictly speaking the case in Fig. A.7, which shows that some uncertainty remains on the derived active doping concentration (especially for CVD13 where only three data points are available for each curve), besides the assumption of bulk mobility. This small uncertainty is believed not to have any impact on the conclusions of this work.

A.3.3 Implantation Profiles

The IIB matrix described in Table A.4 is a B-implanted matrix with Ge PAI under different conditions and laser annealed three times under five different annealing temperatures ranging from 1,150 to 1,300 °C. All layers were implanted in 300 mm wafers with an Applied Materials Quantum X low-energy implanter. The laser annealing was run with an Astra Vantage DSA system from Applied Materials using a diode bar laser operating at 808 nm. The dwell time was 0.5 ms and the chuck temperature 450 °C.

An extra lithographic step allowed us to keep a region of the wafer free of the B implant and hence measure the *substrate* measurement, as defined in Sect. A.1.

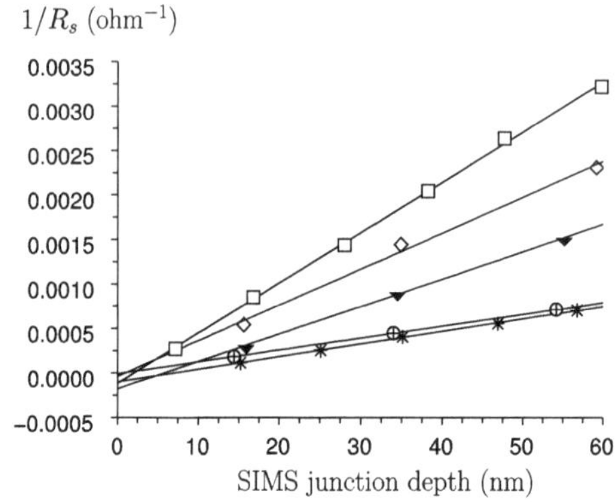

⊕ CVD13 3.x : $1/R_s = 1.33 \times 10^{-5} X_j - 9.64 \times 10^{-6} \Rightarrow P_{\text{doping}} = 1.2 \times 10^{19}\ \text{cm}^{-3}$

✱ CVD12 2.x : $1/R_s = 1.41 \times 10^{-5} X_j - 1.00 \times 10^{-4} \Rightarrow P_{\text{doping}} = 1.3 \times 10^{19}\ \text{cm}^{-3}$

▾ CVD13 2.x : $1/R_s = 3.08 \times 10^{-5} X_j - 1.81 \times 10^{-4} \Rightarrow P_{\text{doping}} = 3.3 \times 10^{19}\ \text{cm}^{-3}$

◇ CVD13 1.x : $1/R_s = 4.42 \times 10^{-5} X_j - 3.33 \times 10^{-4} \Rightarrow P_{\text{doping}} = 4.5 \times 10^{19}\ \text{cm}^{-3}$

□ CVD12 1.x : $1/R_s = 5.64 \times 10^{-5} X_j - 1.19 \times 10^{-4} \Rightarrow P_{\text{doping}} = 6.4 \times 10^{19}\ \text{cm}^{-3}$

Fig. A.7 Derivation of the active doping concentration of CVD layers. The slope of the inverse of the sheet resistance versus the SIMS junction depth gives the conductivity of the layer, which can be translated into active doping concentration, assuming bulk mobility of the majority carriers

A.3.3.1 SIMS Profiles of the IIB Matrix

Figure A.8 shows the SIMS profiles of 8 layers of the IIB matrix. For completeness, the active SIMS concentrations, derived from sheet resistance measurements (FPP) are also indicated by the horizontal lines. For each figure, it can clearly be observed that the sample with a higher annealing temperature, IIB-DxT5 ($x = 03, 05, 07, 13$) is deeper and more highly actively doped than IIB-DxT3. Annealing indeed induces both activation and diffusion. The comparison of the depth of the moderately doped regions ($\approx 10^{18}$ cm^{-3}) in Figs. A.8a versus b–d clearly highlights the impact of the Ge preamorphization (reduction in the channeling tail).

A.4 Impact of Annealing and Preamorphization

As noted in Sect. 7.2.3 and in the conclusions of this work, some effort is still needed so as to quantitatively understand the behavior of ΔR_{ac} on preamorphized and annealed substrates. In this Appendix, we show some preliminary data.

Let us start by investigating the impact of a preamorphizing implant after anneal. This issue has been studied in detail in Ref. [7]. In summary, the behavior

Table A.4 Description of the IIB matrix

	D02	D03	D04	D05	D06	D07	D08	D09	D10	D11	D12	D13	D14	D15
IIB matirx														
n-type monitor substrate	X	X	X	X	X	X	X	X	X	X	X	X	X	X
Ge preamorphizing implant:														
5 keV, 10^{14} cm^{-2}			X	X										
5 keV, 5×10^{14} cm^{-2}					X	X								
10 keV, 10^{14} cm^{-2}							X	X						
10 keV, 5×10^{14} cm^{-2}									X	X				
20 keV, 10^{14} cm^{-2}											X	X		
20 keV, 5×10^{14} cm^{-2}													X	X
B implant:														
0.5 keV, 5×10^{14} cm^{-2}	X		X		X		X		X		X		X	
0.5 keV, 10^{15} cm^{-2}		X		X		X		X		X		X		X
Laser Anneal (3 scans)	X	X	X	X	X	X	X	X	X	X	X	X	X	X
T1=1,150, T2=1,200, T3=1,225°C, T4=1,250,T5=1,300°C														

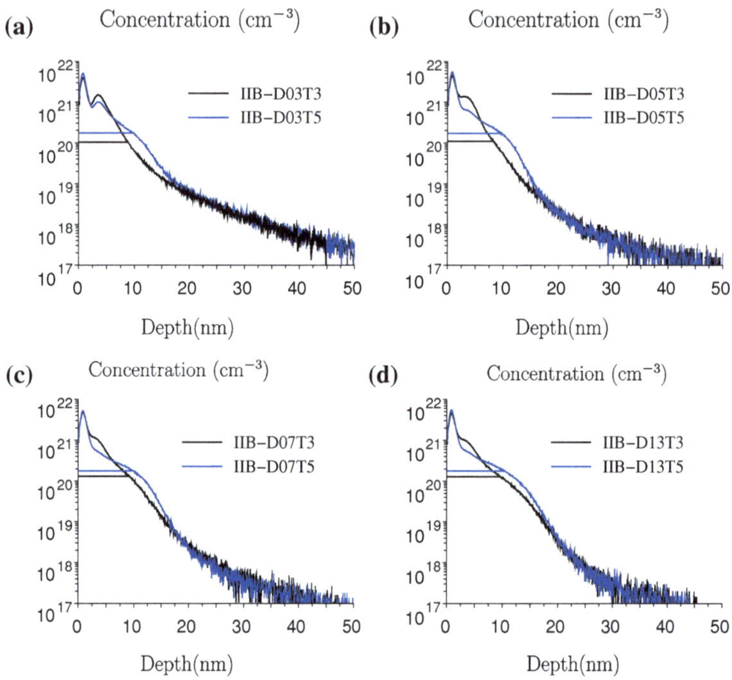

Fig. A.8 SIMS profiles of samples **a** IIB-D03T3 (*black*) and IIB-D03T5 (*blue*), **b** IIB-D05T3 (*black*) and IIB-D05T5 (*blue*), **c** IIB-D07T3 (*black*) and IIB-D07T5 (*blue*),**d** IIB-D13T3 (*black*) and IIB-D013T5 (*blue*). The active SIMS concentrations derived from sheet resistance are indicated by the *horizontal lines*

of the AC reflectance on annealed preamorphized samples can be understood by means of a generalization of its behavior on as-implanted samples, where ΔR_{ac} correlates with the defect density generated during the implantation process [8]. The situation of annealed preamorphized substrates is obviously more complex since, during the anneal, the defects tend to gather in the end-of-range plane. A preliminary analysis has proven that ΔR_{ac} still correlates, at least qualitatively, with the defect density in the end-of-range plane [7]. This is in agreement with the exemplary ΔR_{ac} offset curves shown in Fig. A.9a and b. These figures show the substrate measurements (as defined in Sect. 7.2.2, i.e. no doped layer) measured on four samples with different Ge PAI conditions and laser-annealed three times at 1,300 °C: IIB-D03T5 (no PAI), IIB-D05T5 (5 keV, 10^{14} cm^{-2}), IIB-D07T5 (5 keV, 5×10^{14} cm^{-2}) and IIB-D13T5 (20 keV, 10^{14} cm^{-2}). It is observed that when the PAI energy or dose increases, the amplitude and phase of ΔR_{ac} drop, as a consequence of the reduction in the plasma component and simultaneous increase in the thermal component.

Opposite variations are observed when the annealing temperature is increased and/or the number of laser scans is increased. This is illustrated in Fig. A.9c and d

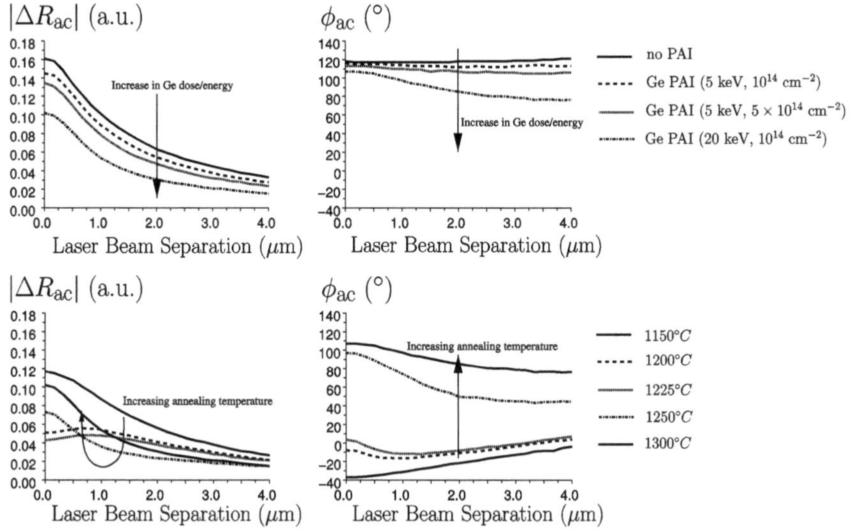

Fig. A.9 Offset curves of **a** the amplitude and **b** phase of ΔR_{ac} as measured on laser-annealed (3 scans, $1,300\,^{\circ}$C) substrates preamorphized with the different conditions indicated on the right hand-side. The amplitude and phase are observed to drop with PAI energy and dose, as a result of the decrease in plasma component and increase in thermal component. The arrows indicate the trend when increasing the Ge dose or energy. Offset curves of **c** the amplitude and **d** phase of the ΔR_{ac} measured on a preamorphized substrate (20 keV, 10^{14} cm^{-2}) for different laser-annealing temperatures. The phase increases monotonically with annealing temperature while the amplitude passes by a minimum. This is the result of the competition between the plasma component, which increases with rising annealing temperature, and the thermal component which decreases with increasing annealing temperature. The *arrows* indicate the trend when increasing the annealing temperature

where the substrate measurements of sample IIB-D13 for different annealing temperatures are shown. It can be seen that a rising annealing temperature leads to a monotonically increasing phase and an amplitude which first decreases before rising again. This behavior can be understood as the result of the competition between the plasma component, which increases with rising annealing temperature, and the thermal component which decreases with increasing annealing temperature. In other words, when increasing the annealing temperature, the signal undergoes a transition from a thermally dominated behavior to a plasma-dominated behavior.

Though these effects are understood qualitatively, they can currently not be modeled quantitatively. This is the reason why we have had to resort to using a linear model with homogeneous recombination lifetime and ambipolar diffusivity in Sect. 7.2.3. We believe this is what has led to some misquantification of the depth of the profiles.

Very interestingly, the impact of annealing on the substrate measurement is also observed on non-preamorphized device-quality substrates. This unexpected

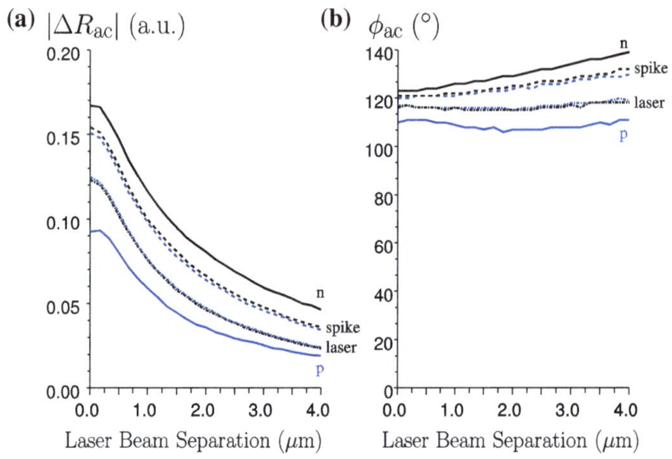

Fig. A.10 Offset *curves* of **a** the amplitude and **b** the phase of ΔR_{ac} as measured on n- (*black*) and p-type (*blue*) device-quality substrates without preamorphization. While ΔR_{ac} depends on the doping type when the device-quality substrates have not been annealed (*full lines*), this dependence disappears after spike-annealing (*dashed lines*) or laser-annealing (*interrupted lines*) these substrates

behavior is illustrated in Fig. A.10, where we compare the impact of a spike anneal (1.5″, 1,035 °C) and a laser anneal (3 scans, 1,300 °C) on n-type and p-type device-quality substrates without any PAI or active implant. It is first surprising to notice that the offset curves measured before any anneal on the n-type and p-type substrates are significantly different. Knowing that device-quality wafers have stringent specificications on the active doping concentration ($<10^{15}$ cm^{-3} for both n- and p-type), this is in disagreement with the experimental data shown in Fig. 6.2 and our model which both expect the same behavior on lowly doped p- and n-type substrates. This difference could be attributed to a different behavior of the surface charging. As noted in Ref [3], this difference has also been observed on monitor-quality substrates, where it strongly decreases when the surface is charged for a longer period. Longer charging measurements on these device-quality substrates should be run to better understand this effect.

As interestingly, after annealing, whether spike or laser, the offset curves on n- and p-type substrates coincide. In other words, while the p- and n-type substrates can be discriminated before annealing, they are equivalent after annealing. Different effects could explain this behavior. First, if the difference observed before annealing indeed proves to be a surface difference, it can then be supposed that the annealing somehow modifies the oxide. Second, it could also be speculated that this behavior is due to the partial activation of the O atoms in these substrates, which act as thermal donors and therefore would make the substrates equivalent after annealing (total oxygen concentration $\approx 10^{17}$ cm^{-3}).

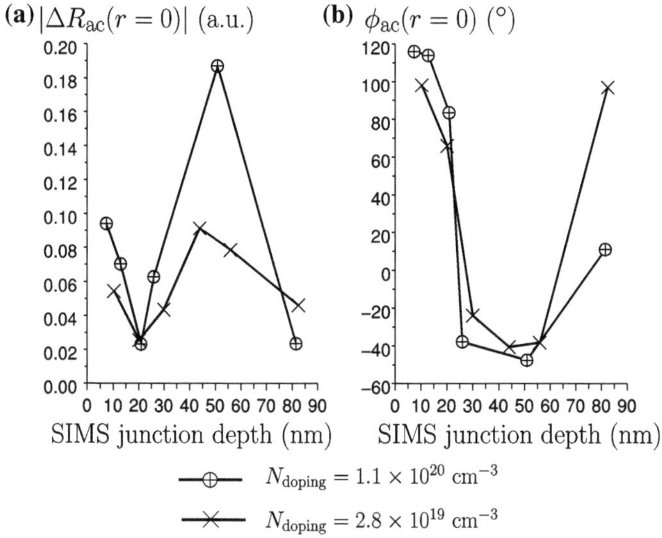

Fig. A.11 Variations with junction depth in **a** the amplitude and **b** the phase of ΔR_{ac} as measured on the CVD layers of the CVD10 matrix. The qualitative behavior of ΔR_{ac} observed on p-type layers in Fig. 6.4b and d is clearly recognized

A.5 N-Type Layers

The CVD10 matrix is a matrix of As-doped CVD layers (n-type). Unfortunately, all these layers have been grown on monitor-quality p-type substrates. They were indeed grown in the early phase of this work when the measurement procedure described in Appendix A.1 and, most particularly, the importance of the quality of the substrates had not been clarified. As a consequence, these layers cannot be discussed quantitatively. In this Appendix, we want to show that n-type layers behave qualitatively like p-type layers, as expected by our model.

We focus the discussion on the values of ΔR_{ac} measured with aligned beams, i.e. we concentrate on the variations in amplitude and phase with active doping concentration and junction depth. This is indeed sufficient to show that n-type layers have a behavior similar to p-type layers. Figure A.11a and b show respectively the variation with junction depth in the amplitude and phase of ΔR_{ac}, as measured on the CVD10 matrix. It can be observed in Fig. A.11a that $|\Delta R_{ac}|$ follows, at least qualitatively, the same behavior as on p-type layers (Fig. 6.4b). In particular, the same rectified cosine behavior is observed. The amplitude of the cosine also increases with increasing doping. The behavior of the phase observed in Fig. A.11b also clearly reminds the behavior of p-type layers (Fig. 6.4d). In particular, the same phase transition is noted when $X_j \approx 22$ nm. This is due to the sign change of the substrate plasma component. In summary, Fig. A.11 clearly indicates that n-type layers behave like p-type layers, as expected by the model we have developed.

For completeness, note that, since all the theory in this work has also been developed in parallel (but not tested) for n-type layers, all the results shown in this work can be extended to n-type layers. To confirm this statement, we have tested the profile reconstruction technique of Sect. 7.2 to As- and P- implanted layers. The obtained profiles for spike-annealed (1.5″, 1,035°) P and As implants (1 keV, 10^{15} cm^{-2}) give acceptable qualitative agreement with the expectations (not shown). The technique has also been tested on laser-annnealed P and As implants (same conditions). However, these implant and anneal conditions result in sub-10 nm deep layers. As a consequence, the observed variations in ΔR_{ac} are of the order of the precision of the technique (low sensitivity).

A.6 Impact of the BGN Electroabsorptive Effect on the Signal Equations

The conclusions of Chap. 3 are that the BGN electroabsorptive effect can be neglected. As a direct consequence, the sinus contributions to Eqs. (2.17), (2.32) have been dropped when writing our final signal Eqs. (6.9), (6.10) and (6.16) . Focusing on the behavior of ΔR_{ac} on box-like profiles, we show here that no BGN electroabsorptive effect is observed in our experimental data, in agreement with our conclusions in Chap. 3.

To prove this statement, let us write the complete equation for ΔR_{ac} on a box-like profile including the perturbation of the extinction coefficient due to BGN. In this case, we have

$$\begin{aligned}\Delta R_{\text{ac}}(r) = {} & \frac{4R_0}{n_0^2-1}\Gamma_0 \exp(-i\theta_0) \\ & \times \Bigg\{ \underbrace{-\beta\left(\frac{1}{m_e}+\frac{1}{m_h}\right)\Delta N_{sub1}\left[\cos(4\pi n_0 X_j/\lambda_{\text{probe}})+\kappa\sin(4\pi n_0 X_j/\lambda_{\text{probe}})\right]}_{\text{substrate plasma component}} \\ & \underbrace{-\beta\left(\frac{1}{m_e}+\frac{1}{m_h}\right)\Delta N_{l1}\left[1-\cos(4\pi n_0 X_j/\lambda_{\text{probe}})-\kappa\sin(4\pi n_0 X_j/\lambda_{\text{probe}})\right]}_{\text{layer plasma component}} \\ & + \underbrace{\delta\Delta T_1}_{\text{thermal component}} \Bigg\},\end{aligned} \tag{A.1a}$$

where $\kappa = (\partial k/\partial N + \partial k/\partial P)/[\beta(1/m_e + 1/m_h)]$ is always positive. The impact of the BGN electroabsorptive effect is therefore a modification in the dependence of the signals upon junction depth. More particularly, the $\cos(4\pi n_0 X_j/\lambda_{\text{probe}})$ dependence is replaced by the modified function $[\cos(4\pi n_0 X_j/\lambda_{\text{probe}})+\kappa\sin(4\pi n_0 X_j/\lambda_{\text{probe}})]$. To illustrate this modified X_j-dependence, Fig. A.12 shows the signal behavior as a function of junction depth for different values of

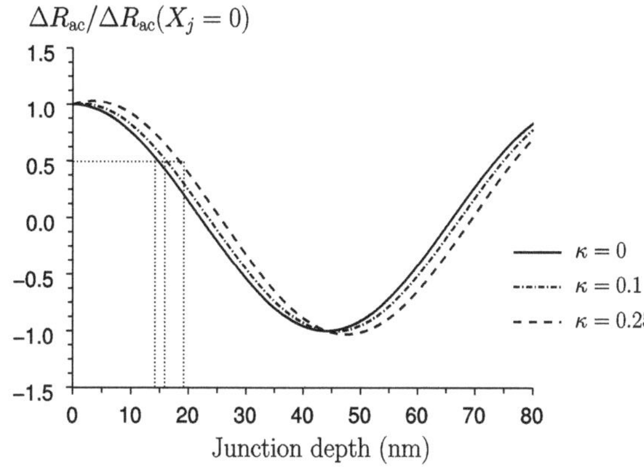

Fig. A.12 Variation in $\Delta R_{ac}/\Delta R_{ac}(X_j = 0)$ as a function of junction depth for three different values of the κ parameter introduced in Eq. (A.1a), assuming negligible layer plasma and thermal components. As shown by the *dotted vertical lines*, when increasing κ, a same value of ΔR_{ac} corresponds to deeper X_j

the coefficient κ. This figure assumes negligible layer plasma and thermal components. The conclusions would be the same if these two components were included.

As highlighted by the vertical and horizontal dotted lines in Fig. A.12, a same value of ΔR_{ac} corresponds to slightly deeper X_j when κ increases. In other words, the consideration of a small BGN electroabsorptive effect would lead our model to deriving slightly deeper values of X_j. Section 7.1.2, however, shows that, with $\kappa = 0$, a very good agreement with SIMS has been obtained. Further, it shows that the derived junction depths might actually be slightly overestimated versus the SIMS junction depths. A shift towards deeper junction depths is therefore to be excluded.

In summary, our experimental data are in agreement with the conclusions drawn in Chap. 3, i.e. the BGN electroabsorptive effect is so small versus the Drude electrorefractive effect that it is not observed experimentally. In particular, based on Fig. A.12, it appears that $\kappa = 0.1$ leads approximately to a 1–2 nm shift in the junction-depth dependence. It can therefore be expected that $\kappa < 0.1$, i.e. $(\partial k/\partial N + \partial k/\partial P) < 0.1 \times \beta(1/m_e + 1/m_h)$.

A.7 Fourier Coefficients of the Auger and SRH Recombination Rates

The transport Eq. (4.54) involve the constant and fundamental modes of the Auger and SRH recombination rates. We here show how to derive their analytical expressions. In order to be general, the electron and hole distributions (Eqs. (4.3) and (4.4)) and the recombination rates (Eq. (4.33c)) are expanded up to their third harmonic. This gives us the opportunity to show by numerical simulations that, as explained in Sect. 4.1.3.3, the generation of second and third harmonics due to the nonlinear (mostly Auger) recombination rate is negligible in TP. In order to simplify the expressions, we also write $N_{cst} = N_{doping} + \Delta N_0$ and $P_{cst} = P_{doping} + \Delta N_0$.

Let us start with Auger recombinations. Equation (4.48) shows that Auger recombinations are a strongly nonlinear phenomenon which involves third powers of the carrier concentrations. Plugging Eqs. (4.3) and (4.4) into Eq. (4.48), the constant Auger recombination rate is

$$\begin{aligned}\mathrm{Rec}_0^{\mathrm{Auger}} &= \mathrm{Rec}_{\mathrm{cst}}^{\mathrm{Auger}} \\ &\quad + 2\left[C_n(2N_{\mathrm{cst}} + P_{\mathrm{cst}}) + C_p(N_{\mathrm{cst}} + 2P_{\mathrm{cst}})\right] \\ &\qquad \times \left(\Delta N_1 \Delta N_1^* + \Delta N_2 \Delta N_2^* + \Delta N_3 \Delta N_3^*\right) \\ &\quad + 6(C_n + C_p)\Re\left(\Delta N_2 \Delta N_1^* \Delta N_1^* + 2\Delta N_2 \Delta N_1 \Delta N_3^*\right),\end{aligned} \tag{A.2}$$

where

$$\mathrm{Rec}_{\mathrm{cst}}^{\mathrm{Auger}} = C_n\left(N_{\mathrm{cst}}^2 P_{\mathrm{cst}} - N_{\mathrm{cst}} n_i^2 \gamma_n \gamma_p\right) + C_p\left(P_{\mathrm{cst}}^2 N_{\mathrm{cst}} - P_{\mathrm{cst}} n_i^2 \gamma_n \gamma_p\right) \tag{A.3}$$

is the Auger recombination rate when only the probe and the constant mode of the pump irradiance are shining. It can be seen that the constant mode of the Auger recombination rate is enhanced under modulated illumination, thus reducing the constant mode of the excess carrier concentration and leaving room for harmonic generation. The fundamental mode of the Auger recombination rate can be written [1]

$$\begin{aligned}\mathrm{Rec}_1^{\mathrm{Auger}} &= \Delta\mathrm{Rec}_{\mathrm{cst}}^{\mathrm{Auger}} \\ &\quad + \Delta N_1\big\{ - \Delta N_1^*\left[C_n(2N_{\mathrm{cst}} + P_{\mathrm{cst}}) + C_p(N_{\mathrm{cst}} + 2P_{\mathrm{cst}})\right] \\ &\quad + 2(C_n + C_p)(\Delta N_1 \Delta N_1^* + 3\Delta N_2 \Delta N_2^* + 3\Delta N_3 \Delta N_3^*)\big\} \\ &\quad + 2\left[C_n(2N_{\mathrm{cst}} + P_{\mathrm{cst}}) + C_p(N_{\mathrm{cst}} + 2P_{\mathrm{cst}})\right](\Delta N_1^* \Delta N_2 + \Delta N_2^* \Delta N_3) \\ &\quad + 3(C_n + C_p)(\Delta N_1^* \Delta N_1^* \Delta N_3 + \Delta N_2 \Delta N_2 \Delta N_3^*),\end{aligned} \tag{A.4}$$

where

$$\begin{aligned}\Delta\mathrm{Rec}_{\mathrm{cst}}^{\mathrm{Auger}} &= C_n \Delta N_1\left[2N_{\mathrm{cst}} P_{\mathrm{cst}} - n_i^2 \gamma_n \gamma_p + N_{\mathrm{cst}}^2 + \Delta N_1^*(2N_{\mathrm{cst}} + P_{\mathrm{cst}}) + \Delta N_1 \Delta N_1^*\right] \\ &\quad + C_p \Delta N_1\left[2N_{\mathrm{cst}} P_{\mathrm{cst}} - n_i^2 \gamma_n \gamma_p + P_{\mathrm{cst}}^2 + \Delta N_1^*(N_{\mathrm{cst}} + 2P_{\mathrm{cst}}) + \Delta N_1 \Delta N_1^*\right]\end{aligned} \tag{A.5}$$

represents the extra Auger recombinations which would occur in time-independent conditions if additional (time-independent) ΔN_1 electron and hole concentrations were added to the already existing N_{cst} and P_{cst}. In TP, $\Delta\mathrm{Rec}_{\mathrm{cst}}^{\mathrm{Auger}}$ is usually larger than $\mathrm{Rec}_1^{\mathrm{Auger}}$, meaning that modulated conditions reduce these extra recombinations.

As for the second and third harmonics of the Auger recombination rate, so as to highlight their generation by the nonlinear Auger process, we write them as

[1] Note that a few constant terms have been added in this expression in order to artifically have $\Delta\mathrm{Rec}_{\mathrm{cst}}^{\mathrm{Auger}}$ to appear

$$\mathrm{Rec}_2^{\mathrm{Auger}} = \frac{\Delta N_2}{\tau_2^{\mathrm{Auger}}} - G_2^{\mathrm{Auger}} \tag{A.6a}$$

$$\mathrm{Rec}_3^{\mathrm{Auger}} = \frac{\Delta N_3}{\tau_3^{\mathrm{Auger}}} - G_3^{\mathrm{Auger}}, \tag{A.6b}$$

where τ_2^{Auger} (resp. τ_3^{Auger}) and G_2^{Auger} (G_3^{Auger}) respectively represent the pseudo Auger recombination lifetime and generation rate of the second (resp. third) harmonic. It can be easily proven that

$$\begin{aligned}\frac{1}{\tau_2^{\mathrm{Auger}}} &= C_n\left(2N_{\mathrm{cst}}P_{\mathrm{cst}} - n_i^2\gamma_n\gamma_p + N_{\mathrm{cst}}^2\right) + C_p\left(2N_{\mathrm{cst}}P_{\mathrm{cst}} - n_i^2\gamma_n\gamma_p + P_{\mathrm{cst}}^2\right)\\ &\quad + 3(C_n + C_p)\left(2\Delta N_1\Delta N_1^* + \Delta N_2\Delta N_2^* + 2\Delta N_3\Delta N_3^*\right)\end{aligned} \tag{A.7a}$$

$$\begin{aligned}\frac{1}{\tau_3^{\mathrm{Auger}}} &= C_n\left(2N_{\mathrm{cst}}P_{\mathrm{cst}} - n_i^2\gamma_n\gamma_p + N_{\mathrm{cst}}^2\right) + C_p\left(2N_{\mathrm{cst}}P_{\mathrm{cst}} - n_i^2\gamma_n\gamma_p + P_{\mathrm{cst}}^2\right)\\ &\quad + 3(C_n + C_p)\left(2\Delta N_1\Delta N_1^* + 2\Delta N_2\Delta N_2^* + \Delta N_3\Delta N_3^*\right)\end{aligned} \tag{A.7b}$$

$$\begin{aligned}G_2^{\mathrm{Auger}} &= -\left[C_n(2N_{\mathrm{cst}} + P_{\mathrm{cst}}) + C_p(N_{\mathrm{cst}} + 2P_{\mathrm{cst}})\right](\Delta N_1\Delta N_1 + 2\Delta N_1^*\Delta N_3)\\ &\quad - 6(C_n + C_p)\Delta N_1\Delta N_2^*\Delta N_3\end{aligned} \tag{A.7c}$$

$$\begin{aligned}G_3^{\mathrm{Auger}} &= -2\left[C_n(2N_{\mathrm{cst}} + P_{\mathrm{cst}}) + C_p(N_{\mathrm{cst}} + 2P_{\mathrm{cst}})\right]\Delta N_1\Delta N_2\\ &\quad - (C_n + C_p)(3\Delta N_1^*\Delta N_2\Delta N_2 + \Delta N_1\Delta N_1\Delta N_1)\end{aligned} \tag{A.7d}$$

Let us move on to the SRH recombination rate. First, we linearize the denominator of Eq. (4.44). In other words, the nonlinearity induced by the denominator of the SRH recombination rate is only taken in first order. The denominator is expanded as follows

$$\begin{aligned}\frac{1}{\tau_p(N + n_i\gamma_n) + \tau_n(P + n_i\gamma_p)} \approx \mathrm{Den}_0\big\{1 &- \mathrm{Den}_0(\tau_p + \tau_n)\\ &\big[(\Delta N_1\exp(i\omega t) + \Delta N_1^*\exp(-i\omega t)\\ &\Delta N_2\exp(2i\omega t) + \Delta N_2^*\exp(-2i\omega t)\\ &\Delta N_3\exp(3i\omega t) + \Delta N_3^*\exp(-3i\omega t)\big]\big\},\end{aligned} \tag{A.8}$$

where

$$\mathrm{Den}_0 = \frac{1}{\tau_p(N_{\mathrm{cst}} + n_i\gamma_n) + \tau_n(P_{\mathrm{cst}} + n_i\gamma_p)}. \tag{A.9}$$

In the context of our investigations, this assumption proves to be good since the constant mode of the carrier distribution (doping *and* constant mode of the excess carrier distribution) is always larger than the other components, whether at low or high injection.

Expanding the complete Eq. (4.44), the SRH recombination rate reads

$$\begin{aligned}\mathrm{Rec}_0^{\mathrm{SRH}} &= \mathrm{Rec}_{\mathrm{cst}}^{\mathrm{SRH}} \\ &- \mathrm{Den}_0^2(\tau_n+\tau_p)\big[2(P_{\mathrm{cst}}+N_{\mathrm{cst}})(\Delta N_1\Delta N_1^*+\Delta N_2\Delta N_2^*+\Delta N_3\Delta N_3^*) \\ &+ 6\Re\big(2\Delta N_1\Delta N_2\Delta N_3^*+\Delta N_1^2\Delta N_2^*\big)\big] \\ &+ \mathrm{Den}_0(2\Delta N_1\Delta N_1^*+2\Delta N_3\Delta N_3^*+2\Delta N_2\Delta N_2^*),\end{aligned} \tag{A.10}$$

where

$$\mathrm{Rec}_{\mathrm{cst}}^{\mathrm{SRH}} = \mathrm{Den}_0(N_{\mathrm{cst}}P_{\mathrm{cst}}-n_i^2\gamma_n\gamma_p) \tag{A.11}$$

is the SRH recombination rate when only the probe and the constant mode of the pump irradiance are shining. It can be seen that more constant recombinations occur under harmonic excitation than in time-independent conditions. This leaves room for harmonic generation.

The fundamental mode of the SRH recombination rate is [2]

$$\begin{aligned}\mathrm{Rec}_1^{\mathrm{SRH}} &= \Delta\mathrm{Rec}_{\mathrm{cst}}^{\mathrm{SRH}} \\ &- \mathrm{Den}_0^2(\tau_n+\tau_p)\big[(P_{\mathrm{cst}}+N_{\mathrm{cst}})(-\Delta N_1^*\Delta N_1+2\Delta N_2^*\Delta N_3+2\Delta N_1^*\Delta N_2) \\ &+ 6\Delta N_1\Delta N_3\Delta N_3^*+3\Delta N_2^2\Delta N_3^*+3(\Delta N_1^*)^2\Delta N_3 \\ &+ 6\Delta N_1\Delta N_2\Delta N_2^*+2\Delta N_1^2\Delta N_1^*+N_{\mathrm{cst}}P_{\mathrm{cst}}\Delta N_1\big] \\ &+ \mathrm{Den}_0(-\Delta N_1\Delta N_1^*+n_i^2\gamma_n\gamma_p+2\Delta N_1^*\Delta N_2+2\Delta N_2^*\Delta N_3),\end{aligned} \tag{A.12}$$

where

$$\begin{aligned}\Delta\mathrm{Rec}_{\mathrm{cst}}^{\mathrm{SRH}} &= -\mathrm{Den}_0^2(\tau_n+\tau_p)\Delta N_1\big[(P_{\mathrm{cst}}+N_{\mathrm{cst}})\Delta N_1^*+\Delta N_1\Delta N_1^*-n_i^2\gamma_n\gamma_p\big] \\ &+ \mathrm{Den}_0\big[(P_{\mathrm{cst}}+N_{\mathrm{cst}})\Delta N_1+\Delta N_1\Delta N_1^*-n_i^2\gamma_n\gamma_p\big]\end{aligned} \tag{A.13}$$

Similar to Auger recombinations, the second and third harmonics of the SRH recombination rate are written as

$$\mathrm{Rec}_2^{\mathrm{SRH}} = \frac{\Delta N_2}{\tau_2^{\mathrm{SRH}}} - G_2^{\mathrm{SRH}} \tag{A.14a}$$

[2] Similar to Auger recombinations, a few constant terms have been added in this expression in order to artifically have $\Delta\mathrm{Rec}_{\mathrm{cst}}^{\mathrm{Auger}}$ to appear. They however disappear in the actual expression.

209

$$\mathrm{Rec}_3^{\mathrm{SRH}} = \frac{\Delta N_3}{\tau_3^{\mathrm{SRH}}} - G_3^{\mathrm{SRH}}, \tag{A.14b}$$

where τ_2^{SRH} (resp. τ_3^{SRH}) and G_2^{SRH} (G_3^{SRH}) respectively represent the pseudo SRH recombination lifetime and generation rate of the second (resp. third) harmonic. It can be easily proven that

$$\begin{aligned}\frac{1}{\tau_2^{\mathrm{SRH}}} = &-\mathrm{Den}_0^2(\tau_n + \tau_p)(N_{\mathrm{cst}}P_{\mathrm{cst}} - n_i^2\gamma_n\gamma_p + 6\Delta N_1\Delta N_1^* + 3\Delta N_2\Delta N_2^* + 6\Delta N_3\Delta N_3^*) \\ &+ \mathrm{Den}_0(P_{\mathrm{cst}} + N_{\mathrm{cst}})\end{aligned} \tag{A.15a}$$

$$\begin{aligned}\frac{1}{\tau_3^{\mathrm{SRH}}} = &-\mathrm{Den}_0^2(\tau_n + \tau_p)(N_{\mathrm{cst}}P_{\mathrm{cst}} - n_i^2\gamma_n\gamma_p + 6\Delta N_1\Delta N_1^* + 6\Delta N_2\Delta N_2^* + 3\Delta N_3\Delta N_3^*) \\ &+ \mathrm{Den}_0(P_{\mathrm{cst}} + N_{\mathrm{cst}})\end{aligned} \tag{A.15b}$$

$$\begin{aligned}G_2^{\mathrm{SRH}} = &-\mathrm{Den}_0^2(\tau_n + \tau_p)[(N_{\mathrm{cst}} + P_{\mathrm{cst}})(2\Delta N_1^*\Delta N_3 + \Delta N_1^2) \\ &+ 6\Delta N_1\Delta N_2^*\Delta N_3 + 6\Delta N_1\Delta N_1^*\Delta N_2] \\ &+ \mathrm{Den}_0(2\Delta N_1^*\Delta N_3 + \Delta N_1^2)\end{aligned} \tag{A.15c}$$

$$\begin{aligned}G_3^{\mathrm{SRH}} = &-\mathrm{Den}_0^2(\tau_n + \tau_p)[2(N_{\mathrm{cst}} + P_{\mathrm{cst}})\Delta N_1\Delta N_2 + 3\Delta N_1^*\Delta N_2^2 + \Delta N_1^3] \\ &+ 2\mathrm{Den}_0\Delta N_1\Delta N_2\end{aligned} \tag{A.15d}$$

A.7.1 Harmonic Generation due to Auger Recombinations

To confirm that the harmonics generated by the nonlinear Auger recombination rate are of negligible amplitude, we have implemented and solved the system of four coupled equations solving for the constant mode, fundamental mode, second and third harmonics of the excess carrier distribution in a homogeneously doped silicon substrate.

The obtained amplitudes are shown as a function of p-type doping concentration in Fig. A.13 ($D_{a1} = 0$ and no BGN are assumed). It can be seen that the second and third harmonics are smaller than 1 % of the constant and fundamental modes, independently from doping. Similarly, the error made on the constant and fundamental modes of the excess carrier distribution when neglecting the harmonic generation of the nonlinear recombination rate is also negligible.

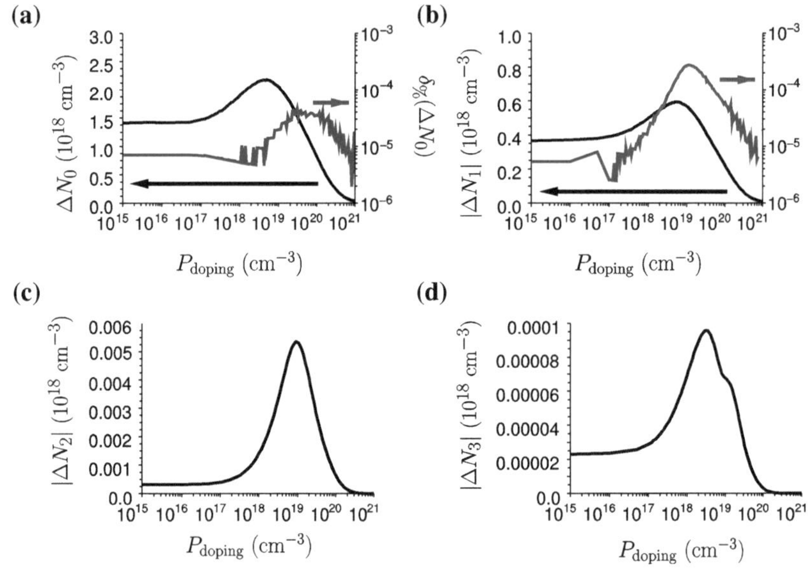

Fig. A.13 Variation with doping in **a** the constant mode, **b** the amplitude of the fundamental mode,**c** the amplitude of the second harmonic and **d** the amplitude of the third harmonic of the excess carrier distribution, solutions of Eq. (4.3.2). These calculations take consideration of the second and third harmonic generation of the Auger and SRH recombination process (Eqs. (A.2)–(A.15a)] but neglect the effects of D_{a1} and BGN. The *blue lines* (*right* scales) underline the relative error made on the constant and fundamental modes when the second and third harmonics are neglected

References

1. J. Opsal, M.W. Taylor, W.L. Smith, A. Rosencwaig, Temporal behavior of modulated optical reflectance in silicon. J. Appl. Phys. **61**(1), 240–248 (1987)
2. F. Dortu, Low-Frequency Modulated Optical Reflectance for the One-Dimensional Characterization of Ultra-Shallow Junctions. Ph.D. thesis, Katholieke Universiteit Leuven, 2009
3. J. Bogdanowicz, F. Dortu, T. Clarysse, W. Vandervorst, E. Rosseel, N.D. Nguyen, D. Shaughnessy, A. Salnik, L. Nicolaides, Nondestructive extraction of junction depths of active doping profiles from photomodulated optical reflectance offset curves. J. Vac. Sci. Technol. B **28**(1), C1C1–C1C7 (2010)
4. T. Clarysse, F. Dortu, D. Vanhaeren, I. Hoflijk, L. Geenen, T. Janssens, R. Loo, W. Vandervorst, B. J. Pawlak, V. Ouzeaud, C. Defranoux, V.N. Faifer, M.I. Current, Accurate electrical activation characterization of cmos ultra-shallow profiles. Mater. Sci. Eng. B-Solid State Mater. Adv. Technol. **114–115**, 166–173 (2004)
5. D.B.M. Klaassen, A unified mobility model for device simulation. 1. model-equations and concentration-dependence. Solid-State Electron. **35**(7), 953–959 (1992)
6. S.M. Sze, *Physics of Semiconductor Devices*, 3rd edn. (Wiley-Interscience, New York, 2006)
7. E. Rosseel, J. Bogdanowicz, T. Clarysse, W. Vandervorst, A. Salnik, Sang-Hyun Han, L. Nicolaides. Study of sub-melt laser damage annealing using therma-probe, in *17th International Conference on Advanced Thermal Processing of, Semiconductors*, 2009

8. W. Vandervorst, T. Clarysse, B. Brijs, R. Loo, Y. Peytier, B.J. Pawlak, E. Budiarto, P. Borden, Carrier illumination as a tool to probe implant dose and electrical activation. in *2003 International Conference on Characterization and Metrology for ULSI Technology* ed. by D.G. Seiler, A.C. Diebold, T.J. Shaffner, R. McDonald, S. Zollner, R.P. Khosla, E.M. Secula, vol. 683 (AIP, 2003) pp. 758–763, http://link.aip.org/link/?APC/683/758/1

MIX
Papier aus verantwortungsvollen Quellen
Paper from responsible sources
FSC® C105338

If you have any concerns about our products,
you can contact us on
ProductSafety@springernature.com

In case Publisher is established outside the EU,
the EU authorized representative is:
Springer Nature Customer Service Center GmbH
Europaplatz 3, 69115 Heidelberg, Germany

Printed by Libri Plureos GmbH
in Hamburg, Germany